Photochemical Reactions in Heterochain Polymers

Photochemical Reactions in Heterochain Polymers

Farukh Fatekhovich Niyazi
Irina Savenkova
O. V. Burykina
and
G. E. Zaikov

Nova Science Publishers, Inc.
New York

For permission to use material from this book please contact us:
Telephone 631-231-7269; Fax 631-231-8175
Web Site: http://www.novapublishers.com

Library of Congress Cataloging-in-Publication Data

Photochemical reactions in heterochain polymers / Farukh fatekhovich Niyazi…[Et al.], editors.

p. cm.
Includes index.
ISBN 13: 978-1-60021-657-2 (hardcover)
ISBN 10: 1-60021-657-9 (hardcover)
1. Heterochain polymers. 2. Photodegradation. 3. Polymers--Deterioration. 4. Polymers--Oxidation. 5. composite materials. I. Niyazi, F.
QD382.H48P46 2007
620.1'1228--dc22

2007011008

Published by Nova Science Publishers, Inc. ✢ New York

CONTENTS

PHOTOCHEMICAL REACTIONS IN SOME HETEROCHAIN POLYMERS

F. Niyazi, I. Savenkova, O. V. Burykina and G. E. Zaikov

"I am too clever to understand what I don't"

Georges Walker Bush,The President of United States of America

"It is necessary to live in the way as if you die today,
but you must work as if you are well"

Dmitriy Sergeevich Likhachiav , Academician of the RAS

PREFACE

Photochemical reactions are the main class of chemical processes. These processes provide the life on the Earth, because due to the photochemical processes, the plants absorb carbon oxide from the atmosphere changing them into the "meal" for all alive creatures on the Earth.

Speaking about photochemistry, it is necessary to mention some famous, well-known scientists. Thanks to their work, the world and marking made a great progress in this field. First of all, it is Winner Nobel Prize prof. Frank Norrish (UK), Norman Allen (UK), Frank Mayo (USA), Herman Mark (USA), Ferens Tudos (Hungary), Georges Geuskenes (Belgium), Georges Smets (Belgium), James Guillett (Canada), Keith Ingold (Canada), Glenn Russel (USA), Georges Hammond (USA) and A.V. Terenin (USSR).

Together with "good" photochemical processes there are some "bad". They destroy every thing. First, let's think about people who get suhburnt. These people destroy their body their organism tries to protect and makes brown pigments which prevent the sun rays to come into the body.

The same processes we can see in photodestruction and simplier chemical reactions. If there is photodestructions, it is necessary to look for photostabilizers.These book (photodestruktions and lightstabilizations) is devoted to these processes. As for examples, we took heterochein polymer. They are the main class of largetonnage polymers, which are used for manning a great variety of polymer and composite materials (including nanocomposites).

Speaking about the processes of destructions and stabilizations of polymers it is necessary to mention the following names: Eli Pearce (USA), Charles Overberger (USA), Nikolai N. Semenov (USSR), Nikolai M. Emanuel (USSR), Moisei B. Neiman (USSR), Norman Grassie (UK), Gerald Scott (UK), Ian MacNeill (UK), Berta M. Kovarskaya (USSR), Nikolai A. Plate (Russia), Viktor A. Kabanov (Russia), Vasilii V. Koshak (USSR), Bob Howell (USA), William Herbert Starnes (USA), Edward Grzywa (Poland), Hans-Joachim Radusch (Germany), Viktor Manuel de Matos Lobo (Portugal) and Alfonso Jimenez (Spain). This lists of names are not complete. We should use 500 names from 30 countries.

We hope that these book will be interesting for the readers and we'll wait for your advice and remarks.

Prof. Niyazi Farukh Fetekhovich
(Kursk State University, Kursk, Russia)
Dr. Savenkova Irina Vladimirovna
(Kursk State University, Kursk, Russia)
Pr. Burykina Oksana Vladimirovna
(Kursk State University, Kursk, Russia)
Prof. Zaikov Gennady Efremovich
(N.M. Emanuel Institute of Biochemical Physics, Moskow, Russia).

Chapter 1

STABILIZATION AND MODIFICATION OF POLYCAPROAMIDE

1.1. PHOTO- AND THERMAL DESTRUCTION OF ALIPHATIC POLYAMIDES

Aliphatic polyamides (PA) and products on their base are polymers with low stability to ultra-violet irradiation and under the effect of solar radiation they destruct less than in a year [1]. In natural conditions ultra-violet part of solar spectrum with wave-length 290-350 nm is more dangerous for these polymers, though aliphatic PAs in this field have very weak absorption of chromophore amide group [2, 3]. That is why while examining the problem of PA phototransformations, unlike other polymers, both possibility of light absorption by impurities and self-absorption should be taken into consideration [4].

As it is seen from the work [5] long-wavelength absorption for extra free amides is characterized by low coefficient of extinction $\varepsilon=0.02$ *l/mole* and by maximum of absorption $\lambda=300$ *nm.* Besides, more long-wavelength absorption, caused by impurities, is observed in solid PA in comparison with model compounds; as impurities there may be products of PA oxidation and initial monomers [6]. Hence, theoretically possible absorption of light by amide group is limited by wave length $\lambda=350\text{-}360nm$, that is photo-transformation of PA may be only sensitized and initiated by impurities, products of oxidation and special additives, in natural conditions PA destructs during 2-3 months and this allows to come to a conclusion that there is possibility for both direct and sensitized transformation of PA under sunlight action.

Many works are devoted to investigation of aliphatic PA light ageing. However, interpretation of obtained results was still contradictory and, as a rule, all investigations come to accumulation of quantitative information about phenomenon of PA light ageing and differences in carrying out experiments (time of irradiation, size of samples, spectral composition of active light and so on) were leading, in their turn, to changing not only relation between product of oxidation, but also composition of products. For example, presence of carbon dioxide and hydrogen and also hydrocarbon in products of oxidation was observed only in some cases. So, formation of hydrogen, characteristic for conditions of photolysis [7] was only lately explained by diffusion restriction of oxygen penetration into polymer [8], and, in its turn, CO_2 content increased in the presence of oxygen, but H_2 was not found, as there is great difference in PA oxidation in the air or in oxygen atmosphere. Investigation of thermal and photooxidative destruction of polyamide fabric in the presence or absence of nitrogen oxides [9] is worth of paying attention to. Analyses of products at ultra-violet light irradiation of ε - caprolactam in nitrogen oxide atmosphere in the absence of oxygen shows presence of α-nitrocompounds, from which hypnitric acid splits out forming caprolidenimide (hydrolyzed further intro amide ε-oxocapronic acid identified by 2,4-dinitrophenylhydrazone) and ε - oxocapronic acid and its ester. Fabric irradiation in the atmosphere with very large concentration of nitrogen oxides leads to the formation of aldehydes, not found in the absence of nitrogen oxides. This shows that photolithic destruction of PA in the atmosphere of nitrogen oxides is connected additionally with reactions of splitting.

Mechanisms of aliphatic PA photooxidation under light action in the conditions of amide group absorption [4] where studied. These reactions way flow under the action of short-wave light which boundary at the Earth's surface is 290 nm. The main result of investigations was discovery of the role of radical – *CONHCHCH₂ – (R₁*)*, being the precursor of all basic products of alkylamide oxidation, for example, *RC*O* and *P₁**, formed at primary photochemical process of photodissociation of amide bond according to the reaction:

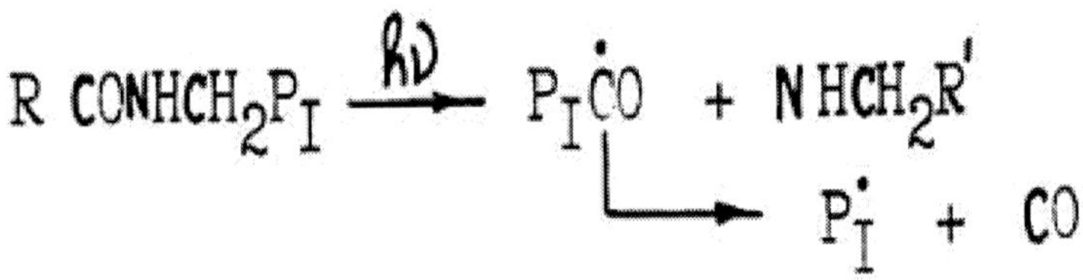

Photolysis of PA-6 in solid state by light 253,7 nm is considered in works [7,8]; gas composition of products is also investigated and this shows the presence of CO-groups.

Analysis of the products of nonsensibilized oxidation of PA and model amides proved special role of radical P_1* which, being formed by stage of H breaking from α- methylene group, led to further autooxidation according to a following scheme [10]:

$$P_1H \xrightarrow{Wo} P_1^* \quad (1)$$
$$P_1^* + O_2 \rightarrow P_1O_2^* \quad (2)$$
$$P_1O_2^* + P_1H \rightarrow P_1OOH + P_1^* \quad (3)$$
$$P_1OOH \xrightarrow{hv} P_1O^* + O^*H \quad (4)$$

where W_0 – the rate of primary initiation; reaction (1) –reaction of initiation; (2-4)-reactions of extension and branching of the chain.

It has been supposed that probability of radical decay of P_1OOH is small and photodecay of peroxides does not make considerable contribution to formation of radicals and products of oxidation, and appearance of radicals P_1* and P_1O_2 in PA leads to efficient breakage of macromolecules. Under some conditions there is possibility for quantitative transformation of redicals P_1* and P_1O_2 and formation of P_1* with quantitative yield from radicals of other types.

While studying kinetics of photooxidation of polyundecanamide, substituted at nitrogen atom by heptyl groups by 20 and 33% it has been found that accumulation of hydroperoxides and imides in substituted PA is much higher than in initial one [11]. Maximum of their concentration grows with the increase of substitution degree. According to some authors perspective light-stabilizers of alkyl substituted PA may be additives of reduced character regarding hydroperoxides.

Irradiation of thin PA films in oxygen at the temperature 77K allowed to accumulate radicals P_1* which completely transformed into radicals P_1O_2* during heating up to 213K, and irradiation at 213K allowed to registrate immediately formation of P_1O_2* [12]. However, in this case there appears additional factor, influencing photooxidation, - it is diffusion restrictions for oxygen penetration into solid polymer. That is why light ageing mainly depends on the oxygen dissolved in polymer film. So, emission of hydrogen is observed in diffusion region, but hydrogen was not found in kinetic region of photooxidation and this may be explained by the interaction of hydrogen forerunner (for example, P_1*) with oxygen [13].

Thus, at nonsensibilized photooxidation of PA formation of hydroperoxides does not influence greatly the process of photooxidation and hydroperoxides are not main products; radical P_1* plays the main role as all other radicals may be formed from it, including peroxides. Hence, while stabilizing PA against action of ultra-violet radiation we may use additives which interact not only with peroxides but with other radicals.

1.1.1. Sensibilized Photooxidation of Polyamides

From the practical point of view mechanism of sensibilized PA photooxidation is interesting, as there is possibility to consider mechanism of PA product light ageing , initiated by impurities, products of oxidation and different additives (dyes, pigments, fillers and so on). And this may be done if the process of destruction under the action of long-wave light, which is not absorbed by polymer chromophore groups, will be investigated.

Impurities of mechanical character may get into polymer in the process of its production or processing and also during polymer use. Some authors think that inpurites not only deteriorate the process of polymer processing, but during irradiation in the air the lead to the accumulation of aldehyde groups [14, 15].

In other works [16, 17] influence of impurities of ferrum, aluminium, coper on the rate of PA destruction at initiated photooxidation is not found. Sensibilizing action of ferrum traces and other transition metals in PA photooxidation is found in works [18-20] where it is shown that the destruction process, catalyzed by ferrum impurities, flows slowly when products from PA are stored at room temperature.

Mechanism of photooxidation of aliphatic PA in the conditions when absorption of light by amide group (λmax=360 nm) is excluded and there are no special additives is studied in the work [21]. The authors show that the product, sensibilizing oxidation, is formed under the light action and this product, unlike other polymers, is not hydroperoxide. It is found that the rate of photooxidation is defined by stationary concentration of this product and does not depend on light intensity. The scheme of photooxidation, including quadratic break of the chain, photobranching in semiproduct of dark oxidation and product destruction are given here.

There was undertaken an attempt [12] to show, by methods of luminescence, luminescence centres in PA, intensity of which does not depend on the light intensity, causing photooxidation of polymer. Corrected spectra of PCA fluorescence are characterized by maxima of excitation $\lambda max^{ex} = 360$ nm and emission $\lambda max^{em} = 435$ nm. According to the data of the work [24] fluorescence, characterized by $\lambda max^{ex} = 365$ nm and $\lambda max^{em} = 445$ nm should be classified as fluorescence of impurities – products of PA oxidation and initial monomers.

These data agree quite well with conclusions of the work [22] in which it was noted that change of fluorescence intensity happens simbatly with the change of quaintity of formed C=0 bonds. Similar results were got by Allen and Makk-Kellar [23] while studying fluorescence of model compounds and nylon – 6,6. The authors explain the nature of polymer luminescence by the presence of fragments, being in the main polymer chain, consisting of C=0 groups with conjugated double bond of ethylene type.

Formation of these groups is observed mainly at the stage of polycondensation and also fiber formation [24]. Their identification was carried out according to ultra- violet spectra and spectra of fluorescence, moreover main groups were oligoenimic with the content up to 44 mole/kg.

Interesting results are got in the work [25] in which the authorse suppose that luminescence of PCA may be attributed to the presence of compounds of keto-imide structure in the polymer, these compounds being characterized by $\lambda max^{ex} = 356$ nm and λmax^{em} 417 nm, connecting the formation of these products at the stage of polymerization of ε-caprolactam according to the following scheme:

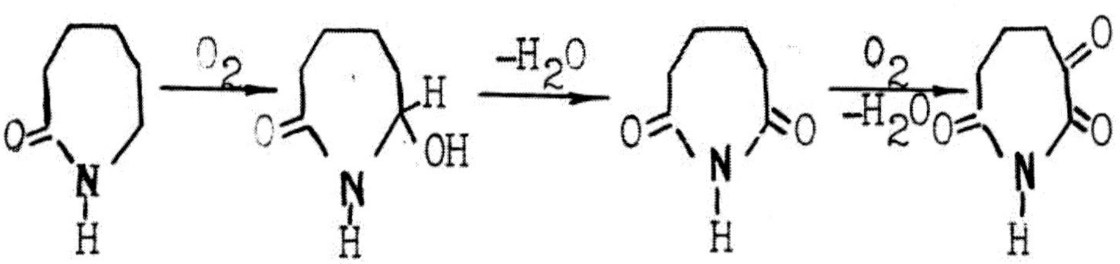

This point of view is shared by Postnikov and his research workers [12, 26], who consider that keto-imide compounds may be formed even at the early stage of PA production. They were the first to construct the simplest scheme of reaction capable to describe mechanism of PA photooxidation taking into account formation and consumption of keto-imide compound:

1. $P_IO_2^{\bullet} \xrightarrow{+PH_2+O_2} P_IO_2^{\bullet}$

2. $P_IO_2 \longrightarrow A + r^{\bullet}$

3. $A \xrightarrow{+2O_2} X$

4. $X + r^{\bullet} \xrightarrow{P_IH} P_IO_2^{\bullet}$

5. $P_IO_2^{\bullet} + r^{\bullet} \xrightarrow{-O_2}$

6. $X \xrightarrow{+O_2} 2P_IO_2^{\bullet}$

where X- keto-imide; A- intermediate product, preceding keto-imide, r*- light radical of HO_2*, O*H type.

Photoinitiator X, primary concentration of which in PA depends on prehistory of samples and exceeds, as a rule, stationary value, achieved under light action, is practically the only source of PA fluorescence.

Suggested mechanism allowed to describe quantitatively PA photooxidation at ultra-violet and long-wave irradiation; there were got equations allowing to determine concentrations of P_1O_2*X [27, 28, 29].

Thus, analysis of literary data allows to come to a conclusion about presence of PA luminescence, related to the product of keto-imide structure, which is formed and consumed in chain processes, performing here the function of photoinitiator. Unlike traditional schemes of photoinitiation intermediate hydroperoxides does not play great role at photooxidative PA destruction.

1.1.2. Photooxidation of Dyed Aliphatic Polyamides

Sensibilized mechanism of PA oxidation in the presence of dyes, pigments and other additives is interesting from the point of view of practical use. Knowledge of this mechanism allows to select effective additives to increase polymer photooxidative stability. Sensibilizing action of TiO_2 is shown in the work [29]. TiO_2 is used for dulling PA fibers, which leads to decrease of physical-mechanical strength of yarn from PA-66 at light action with wave-light λmax=360 nm. Sensibilizing activity of TiO_2 is displayed at generation of singlet oxygen $^1O_2$ and anion-radical O_2^- in the presence of water – radical O*H and HO_2*, which, in their turn, are able to sensibilize PA oxidation. Dyes and pigments may have similar effect.

While investigating some xanthene, thiazine, azo and antraquinone dyes it has been shown that they sensibilize photodestruction of aliphatic PA [30]. At studyng reaction between α, β – nonsaturated carbonye groups of PA-66 and photoactive dye "Dispersed yellow 13" it has been found that the process of triplet-singlet energy transfer takes place at the distance of 4-6nm between donor and acceptor. This process is the reason of low light stability of the dye and its photodestructive effect on polymer [31].

Determination of primary processes is very important for solving problems of dyed fibers photodestruction. Today there are several theories trying to explain photodestruction of dyed PA materials. They suppose that excited molecule of a dye transfers energy to molecular oxygen, forming singlet oxygen $^1O_2$, which, in its turn, causes destruction of polymer according to the following scheme:

$K_p \xrightarrow{h\nu} K_p^*$
$K_p^* + O_2 \rightarrow K_p + {}^1O_2$
${}^1O_2 + P_1H \rightarrow$ products of oxidation (in the absence of water)
${}^1O_2 + H_2O \rightarrow 2H_2O_2$
$H_2O_2 + P_1H \rightarrow$ products of oxidation (in the presence of water)

According to another theory [32] the first stage of destruction is hydrogen atom (or electron) breaking off the polymer by the excited molecule of a dye:

$K_p \xrightarrow{h\nu} K_p$ $\qquad\qquad K_p + O^*H$

$P_1^* + O_2 \rightarrow P_1O_2^*$ (in the absence of water)
$P_1O_2 + K_p \rightarrow P_1OOH + K_p$

Effect of water is connected with the electron transfer from ion OH^- to excited molecule of a dye and, in such a way, forming radical O*H:

$K_p^* + OH^- \rightarrow K_p + O^*H$
$O^*H + P_1H \rightarrow H_2O + P_1^*$ (in the presence of water)

Since polyamides are classified as compounds weakly reacting to O_2, then the second mechanism is more probable for them, though this question is still open because photodestruction of dyed PA depends also on medium, dye properties and conditions for carrying out experiments.

1.1.3. Thermooxidative Destruction of Polyamides

Macromolecule of PA consists of methylene chains connected by amide groups, which cause vulnerability of these polymers at thermooxidative destruction.

Thermooxidation of PA-fibers becomes visible at temperatures 100-150 ^{0}C. PCA - fiber nonreversibly loses 58-65% of initial strength after warming-up in the air at the temperature 180°C for 2 hours, and after 14 hours heating – 75%. PA-fiber loses a great part of molecular weight at high temperatures (270-350 ^{0}C) [33].

That is why, destruction processes go on at temperatures of PA synthesis and processing. And these processes lead to the change of PA structure, decline of phisico-chemical properties of PA-products, which influence their thermostability during usage.

Detailed survey on thermal destruction of a wide range of polymers and PA, specifically, is presented in monograph of Kovarskaya [34]. There are a few works in literature devoted to investigation of the kinetics of oxidation at relatively low temperatures. Kinetics of PCA thermooxidation has been studied in the work [35] having the aim to determine the possibility of forecasting of shelf lives and usage of PA-materials and products on their basis, proceeding from the temperature dependence of their "induction periods of oxidation".

Pakhomov and his assistants [36] note, that loss of strength is defined by kinetic destruction of molecular chains in amorphous regions of PA. Low values of activation energy of thermooxidative destruction may be explained by PCA water receptivity and hydrolysis of amide bonds at the time of warming up. So, for dry PCA-E_{act}=43 Kcal/mole, and in the presence of moisture - E_{act}= (20-30) Kcal/mole. Last conclusions agree with the results of investigation [37] quite of satisfactory.

Kinetic model of thermooxidation of aliphatic PA is given later works [24], model including formation of azomethine structures during the interaction of aldehyde and ketone groups with end NH_2 – groups of PA and further oligomerization of azomethine structures, accompanied by the appearance of conjugated links and PA dyeing. It is noted that calculated values of the change of functional groups (CO; COOH; NH_2) concentrations and also lengths of oligoenimic chain in the process of thermooxidation correlate well with experimental kinetic data on thermooxidation of PA-6 and PA-66 at 160-180 ^{0}C.

Data on thermogravimetric analysis of PA-6 destruction in the air are given in the work [38]. PA loses 25% of mass at warming up in the interval 25-325 ^{0}C, and main transformation (92% of mass) finishes at 470 ^{0}C. The authors of the work [39] have studied composition of gas phase, formed at thermal destruction of PCA using infrared spectroscopy. Comparison of obtained spectra with spectra of model compounds allowed to identify water steams, CO_2, and also MeOH and EtOH in the gas phase.

With the aim to study PA transformations Makarov [40] used method of free radicals initiation by thermal and photochemical decomposition of peroxides. The author succeeded in finding high efficiency of PA macromolecules breakages under the action of free radicals at the temperature 20-98 ^{0}C (within the limits of operational temperatures). He also succeeded in determination of reactions sequence and revealing the phase directly responsible for the acts of polymer chains destruction. It is shown that in the conditions of thermal initiation transformations of peroxides are caused by macroradicals and photochemical – by own radicals of peroxides.

Postnikov and Makarov [41] investigated thermal decomposition of benzoyl in PA at temperature 65-100 ^{0}C in atmosphere of O_2 and in inert atmosphere. They offered the formal scheme of the process allowing to explain all obtained experimental kinetic regularities according to which there takes place initiation without breakage of macromolecule at homolytical decay of $(C_6H_5)_2O_2$ with further breaking of hydrogen atom from polymer with the formation of benzoic acid and macroradical P_1*(CH_2CONHC*HCH_2). These works were probably the only where there was made an attempt to describe quantitatively the process of thermal PA destruction.

Thus, analysis of literary data shows that up to now there is not clear and full idea about the mechanism of PA thermooxidative destruction. And, as a result, there is much indefinite in the question of its stabilization and modification, the more at introducing additives directly into polymer melt [41]. Besides, it should be taken into consideration that majority of these additives possess low thermal stability, and this eliminates their use at the synthesis and production of fibers.

1.1.4. Photostabilization of Polyamides

Need for PA material stabilization arises at the stage of polymer processing, as in this case the polymer, as a rule, should be transferred into flowing state. Stabilization of polymer materials plays a very important role, first of all, to the action of light and heat, oxygen and air.

Interest to this problem is proved by a great number of scientific publications in our country and abroad [1, 2, 8, 10, 42]. However, it is impossible to make systematic analysis of stabilizers efficiency, mechanism of their action, to form scientific approach to the synthesis and use of additives in different and specific cases according to literature surveys [44, 45] and patent information [46, 47], where many classes of compounds are presented as light stabilizers. Interaction of stabilizing efficiency of the additive with its structure is not investigated for many compounds of one class. There are no reliable numerical data on the expediency of use of many organic compounds as ultra-violet absorbers and suppressors of polymer excited states, though necessity in such investigations was repeatedly noted at conferences and symposiums on ageing and stabilization of polymers [48-52].

Photo- and Thermostabilizers of Polyamides

A great number of scientific publications in our country and abroad stress the interest in this problem [2, 42, 10, 46, 53-58]. Metallo-organic stabilizers have found their application for stabilization of polymer materials from PA. In the first place these are compounds of metals with variable valency: copper, lead, zinc and so on. Metal copper, salts of univalent and bivalent copper of inorganic and organic acids [55], mixture of copper with hydrogenised bisphenol and triphenilphosphate [56] from the metal compounds, used as photo – and thermoinhibitors of PA oxidation, are mentioned in major patents. However, these compounds in are badly combined and unevenly distributed in polymer and this makes worse their stabilizing activity because of their low migration in polymer matrix.

Copper macrocyclic complexes of tri – and zoindobenzol in the presence of potassium iodide, known as stabilines [57, 58] (for example, stabiline 10, stabiline A), inhibiting destruction processes and facilitating the formation of more favourable polymer structure are worth of paying attention to. Copper complexes with tiadiazolysoindole and tiadiazolysoindolphenilene in the mix with halogen - , amino – or phosphorus compounds have found their application [59],

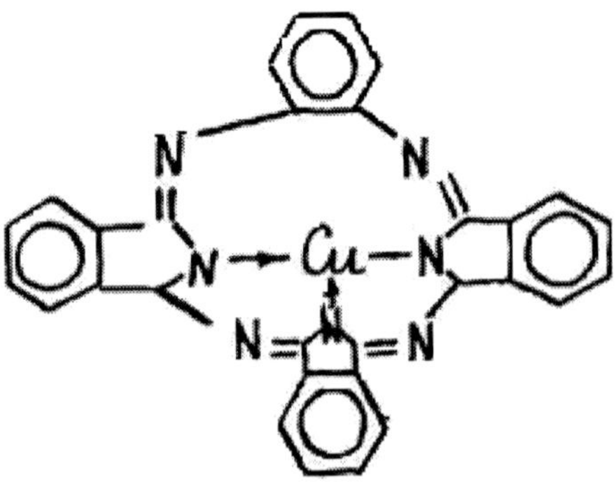

as copper complexes on the basis of polyhexoazocyclanes [60], where it is shown, that introduction of these compounds reduces photo – and thermal destruction of PA at the expence of the presence of copper, halogens, *NH* – groups in stabilizer.

Interesting thermo – and light stabilizer is the product of $CuCl_2$ reaction with 2 – mercaptobenzimidazole [61].

Black, as internal filter regarding ultra-violet light [42], is used with high efficiency. Besides, since black usually contains stable free radicals, it may react with photochemically formed active polymer centres, in such a way reducing the rate of the process of chain continuation [62]. Black is also used as suppressor of triplet and singlet polymer states.

Black's different stabilizing activity is displayed depending on the nature (furnace, channel, thermal, lamp, acetylene black) and the size of the particles.

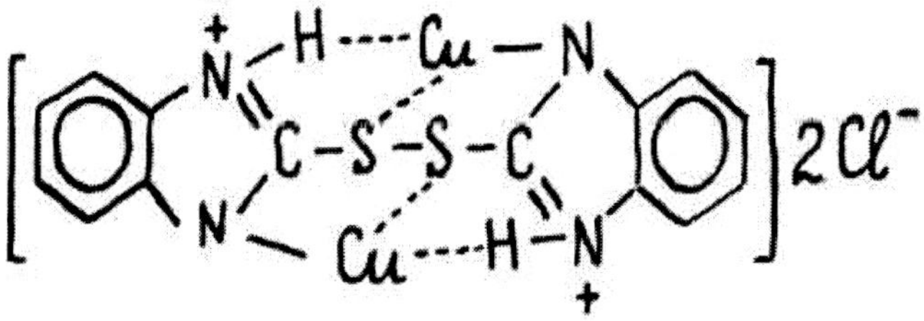

White pigments, such as TiO_2 and ZnO [63] are used with the purpose of ultra-violet rays absorption and for dulling of PA-fibres and films. Titanium oxide begins to reflect light only at the wave-length more than 340 nm. However, TiO_2 action at irradiation of filled polymers is not always identical it may behave as sensibilizer at photooxidation and this requires additional stabilization [64]. For example, processing TiO_2 by small quantities of aluminium, silicon and zinc decreases negative action of pigment.

Zinc oxide is one of more effective and economic white inorganic stabilizers, especially for ultra-violet region from 240 to 380 nm [65], however, its use is limited because of corrosion activity in respect of applied equipment for polymer processing.

Ultra-violet absorbers on the basis of derivatives of 2 – oxybenzophenol: specifically Benzon OA, where R_1=H and R_2 =C_7H_{15}-C_9H_{19},

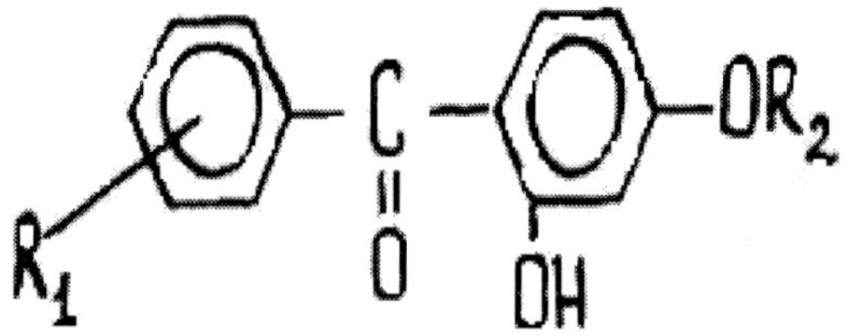

Benzon OM, where R=H and R_2 =CH_3; 2 – oxyphenylbenzotriazoles:

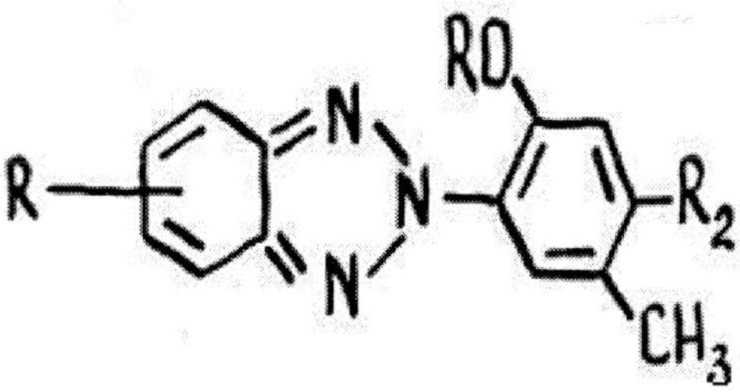

specifically Benazole II (Tinuvine II) at $R_1 R_1$ and R_2 =H; Benazole PXB (Tinuvine 326) at R=H, R_1=C, R_2=C(CH_3) and others; aryl ethers of benzoic, salicylic, tetraphthalic and isophthalic acids [21, 66] are of interest. Mechanism of shielding effect of such stabilizers is in absorption and transformation of short-wave light energy and also in suppression of triplet and singlet excited states of polymers.

Interest in *N, N*[1] - diaryloxamides [67], as the class of non-toxic ultra-violet absorbers possessing light efficiency has raised lately:

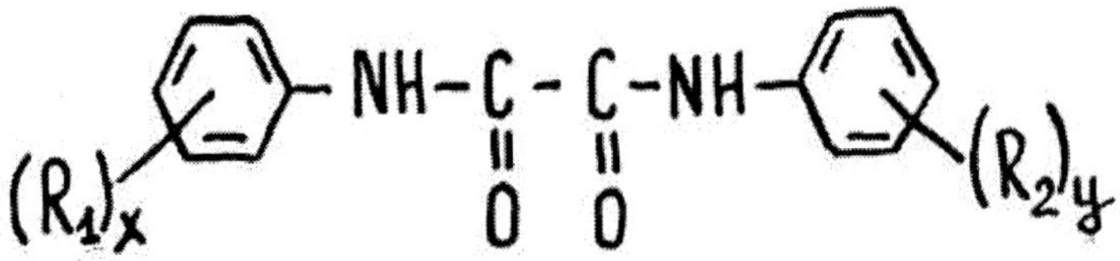

These compounds are similar to such industrial stabilizers as Sanduvor V.5.4 – 2 – ethyl – 2 – ethoxydiphenyloxamide and Sanduvor V.5.4 – 5 – tretbutyl – 2 – etoxy – ethyldiphenyloxamide [68]. Stabilizing effect of such compounds is realized mainly according to physical mechanism of stabilization; reactions of radical inhibition are improbable [69].

Great attention in literature is paid to stabilizers – spatially difficult amines and their derivatives [10, 70] –, which act as inhibitors of radical processes of polymer destruction. Difficult piperidines and their nitroxyl radicals are among them:

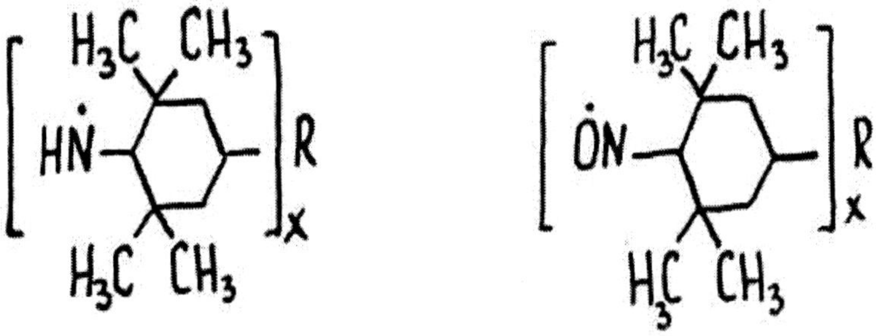

However, they are ineffective in the presence of peroxides and hydroperoxides, which decrease value of these stabilizers at long photooxidation.

Tinuvine 770-bis (2, 2, 6, 6 – tetramenthylpiperidile – 4) sebacate, proposed by the firm "CIBA" [71], and some polymer addititives, including fragments of piperidine cycle are used in industry. As a rule, stabilizers, containing stable radical in their structure, are also inhibitors of thermooxidative destruction.

Additions of amine stabilizers greatly increase thermal stability of PA at long ageing [72]. For example, such antioxidants as *N,N*[1] - di – β – naphthyl – p – phenylene diamine; N,N[1], β – phenylcyclohexil – p – phenylene diamine; N,N[1], β – diphenyl – p – phenylene diamine and phenyl – β – naphthylamine conserve strength of PA – fibres by 70-95% at it warming up for 2 hours at 200 ^{0}C (unstabilized fibre – 22%). Despite high inhibitive activity their use is limited, as intermediate products of their transformations dye polymer.

Good low-dyeing antioxidants are derivatives of 4 – oxidiphenylamine [73, 74], for example, 2, 2 – (diphenylaminophenoxi) diethyl ether, known as stabilizer H-1; 2, 2 (naphthylaminophenoxi) diethyl ether [75]. They are used for thermal stabilization of PCA – threads of technical purpose, though some produced forms of stabilizer H-1 initiate homolytical break of PCA macromolecules and inhibit secondary processes of polymer destruction to a different extent.

There is opinion that stabilizing additives should be colorless [10, 76], on the other hand fluorescent additives (optical bleach) are introduced into polymer in order to compensate yellow painting, which appears at PA processing and also to increase luster and whiteness of products. However, there are works in literature in which there is opinion that it is more preferable to use dyes as stabilizers [77-81], especially if it is necessary to get dyed polymers.

New results are given in the work [77], where it is studied the effect of light spectral composition and dye concentration on the efficiency of photooxidation in PA material, there also studied action of the air oxygen. However, there are practically no data about the effect of dyes and products of their photochemical decay on the polymer itself. Kalontarov and Kharkharov [80] while investigating atmospheric stability of PCA-fibre, dyed by active dyes of dichlortriazine series, have found that after 40-days testing in the atmosphere of Saint Petersburg the given strength exceeds such value for dyed samples by 10-12%.

There has been voiced a supposition [81] that light stabilizing activity of dispersed dyes regarding PA is displayed by the mechanism of PA excited states suppression, but mechanism of ultra-violet shielding does not contribute greatly into the total effect of light protective action. It is also supposed that inhibition of free radical transformations in polymer matrix by dyes may take place [82].

Effect of dyes of different class on light ageing of PA is studied more detailed in the work [83]. It appeared that dyes, used by the authors, sensibilize PA photooxidation according to radical mechanism, however the total action of the dye may be stabilizing owing to shielding effect, and in some cases according to the mechanism of antioxidant action.

Investigation of the effect of the additives of azo dyes on the proceeding of radiation-chemical processes of jointing, destruction, gas emission at irradiation of aliphatic PA in the air and in vacuum has shown that introduction of dyes more effectively reduces the rate of jointing and destruction at irradiation in vacuum, amount of gases at PA radiolysis decreases [84]. According to the authors, protective action of azo dyes is in accepting primary products of destruction and redistribution of absorbed energy between molecules of polymer and dye.

The same results were obtained by the authors of the work [85] who were studying effect of modifier and azo dyes, formed directly in PCA-fibre itself, on photooxidative destruction and photolysis of PCA. In this case azo dyes inhibit both photochemical decay of the fibre at ultra-violet light in vacuum and destruction of dyed fibre.

However, enumerated stabilizers have one vital defect, being expressed in use in thermo-unstable compounds (with some exception), owing to which their use for PCA stabilization is limited according to technological reasons. The task of getting thermostable stabilizers for PA is very topical today. Besides, taking into account the fact, that stabilizing additives are introduced in the amount from 0,25 to 2,0% and only in some cases for coatings up to 10%, it may be supposed that more perspective will be stabilizers acting according to complex physico-chemical mechanism of stabilization including shielding, suppression and inhibition.

One of the most rational methods of PA stabilization may be introducing thermostable stabilizing additives of dyes into polymer mass. In connection with the said above it is necessary to consider modern state of PA mass dyeing.

1.1.5. Polymer Dyeing in Mass

At usual surface dyeing, based on the diffusion process of dye penetration into fibre, inverse process – yield of dye from polymer matrix – is realized in one way or another, and this causes reduction of the products dyeing intensity. But because of structural inhomogeneity of fibres (presence of crystalline and amorphous regions) the dye is distributed in them nonuniformly, resulting in uneven painting and deteriorating colour indexes. Diseconomy of bath dyeing process owing to incomplete sampling of the bath and need in carrying out additional finishing operations should be noted too.

Since there are data on stabilizing activity of dyes at photo – and thermal destruction of PCA [86] introduction of such dyes-stabilizers by the method of mass dyeing, that is adding the dye into polymer at any stage of its production or forming the fibre is of great interest.

Coincidence of polymer synthesis or shaping fibres from them with dyeing is a progressive method in technologycal and economic respect by which fibres in completed marketable state may be directly obtained at chemical fibres plants. It should be noted that mass dyeing, being practically wasteless production, acquires great significance, compared with other methods, in connection with stern measures adopted nowadays in the field of environment protection. Besides, there is no longer any necessity for sewage treatment of textile-finishing manufacture from different pollutions (dye, surface-active agents, salts and so on) because of the service water shortage. The most important is the fact that dyes for mass dyeing in the majority of cases, provide painting with high stability indexes to all actions, often unachievable at usual dyeing methods. Dyes used for mass dyeing do not have a bad effect on fibre strength, but in some cases they are able to protect it from photochemical destruction.

Methods of Mass Dyeing of Polycaproamide

The process of thermoplastic polymers production and shaping is carried out at high temperatures, achieving 300 ^{0}C, which, in its turn, puts strict restrictions to the choice of dyes and their assortment because the most part of the used dyes for PA – fibres dyeing do not withstand long action of high temperature and agressive melt medium. That is why, for PA – fibre dyeing in the process of polymer production and shaping there may be applied only those dyes which meet definite requirements, namely:

(a) to be resistant to high temperatures (250-300 ^{0}C) throughout the whole process of production from the melt; to be resistant to agressive melt medium, chemical action both on the part of separate components, included into the melt, and products of their chemical decay and physical action in the process of polymer processing;
(b) to dissolve and coincide with the polymer, and in the case of pigment – to disperse quite well;
(c) to impart stable to physico-chemical action paint to the fibre.

The dye introduced into reaction mass at the stage of polymer synthesis is in much more rigid conditions. At this time this dye is exposed to the action of monomers which, as a rule, are more reactive that the polymer.

Dyeing composition may be introduced at any stage of polymer synthesis in the form of powder, suspension, paste. For this type of dyeing the stage of paste production, containing up to 30% of organic dyes and not more than 15% of pigments in different carriers (monomers, oligomers and so on). This method is rational and economic while producing large lots of synthetic fibres at large – tonnage continuous machines; however all this prevents the production of wide range of dyed polymers because transition from one colour to another is difficult and this requires technological equipment cleaning. Besides, long process (6-12 hors) leads to the dyes destruction. This is shown in the two-tones of painting throughout the whole technological process and requires application of very thermostable dyes. Methods of PA-fibre dyeing in mass (before polymerization), in granules and in the melt are considered in the work [87]. TiO2, pigments on the basis of *Fe, Cr, Cd* oxides, ultramarine, phthalocyanine pigments are usually introduced before polymerization. PA granules dyeing is carried out by direct mixing with pigment powder in dry and wet form. Metallocomplex and acid dyes in ratio of 1:2 are used for such painting, but some part of the dye gets into waste waters and this leads to economic problems. These defects may be eliminated if the dye is introduced directly into finished polymer mass. Polymer dyeing in mass through the stage of producing polymer concentrate of the dye (PCD) may be very important [88]. In this case methods of dyeing at polymer synthesis and by introducing dyed composition into polymer melt are coincided. Producing polymer concentrate of the dye is fulfilled during the process of polymer synthesis in the presence of large amounts of the dye (up to 50% from polymer mass).

Both initial granulate and ground industrial wastes of polymer may be used according to this method (similar in similar). Dyeing may be carried out according to both periodic and continuous scheme because more uniform distribution of the dye in polymer mass is achieved here and this gives the opportunity to get the best monotone painting of the fibre throughout the whole technological process. However, during this method only very thermostable dyes may be included into the assortment, because the process is carried out at high temperatures (250-280°C) during the long time. But as PCD is the composition of resin-dye with the dyeing component content from 10 to 50%. The role of dye thermal stability is high and it is desirable that the nature of resin in PCD should be the same as of granulate being dyed.

Dyes for Polyamide Dyeing in Mass

Dyes, used for thermoplastic polymers dyeing, may be divided into two groups: dissolved in spinning melt and high-dispersed pigments which must be uniformly distributed in polymer melt and provide stable dispersions. Both groups have their advantages and disadvantages. In the first case high demands are made on dyes solubility and in the second – on the degree of dispersion of dye particles.

Use of dyes does not require fine dispersion and, being dissolved in polymer melt, they as a rule do not influence physico-mechanical properties of the melt and dyed fibre but sometimes improve them. Besides, the amount of dye, necessary to get bright tones of dyed fibre is less than at using pigments. However, dyes are mainly less thermostable and that is why they withstand less time of staying in melt. Light stability of the fibre, painted by these dyes, is also less in most cases which shows feasibility of purposeful search and synthesis of new dyes, especially meant for PA dyeing in mass.

As pigments, which are widely used in manufacture of dyed PA, titanium dioxide – first-rate bleach; coloured mineral pigments, containing black or black with paraffin [89] have found their application; sometimes carbon products of the pyrolysis of organic compound of dymethylizozincomyronaphtha type [90] are used instead of black.

For the last time a great number of new dyes, such as azo dyes, derivatives of antrapyridon and antrapyrimidon [44], derivatives of perylene [91], water-soluble ethers of antrachinon [92] condensation products of symmetric and asymmetric indigoide colours with aryl-cyclohalogenides [93], derivatives of phthalocyanine [94] and others have been synthesized.

Most of the azo dyes destruct because of the low resistance of azo-groups to reducing medium of polymer melt. Azo-dyes metallo-complexes with the composition (1:2), not having free sulfo – and carboxyl groups in the molecule composition are the exception.

Formation of the complex, involving N-atom of azo-group into coordination bond, increases stability of the latter [95-98]. Dyes of different colours may be selected from metallocomplexes, however derivatives of pyrasolon possess the highest stability and dye PA into colours from yellow to red, for example, yellow dye:

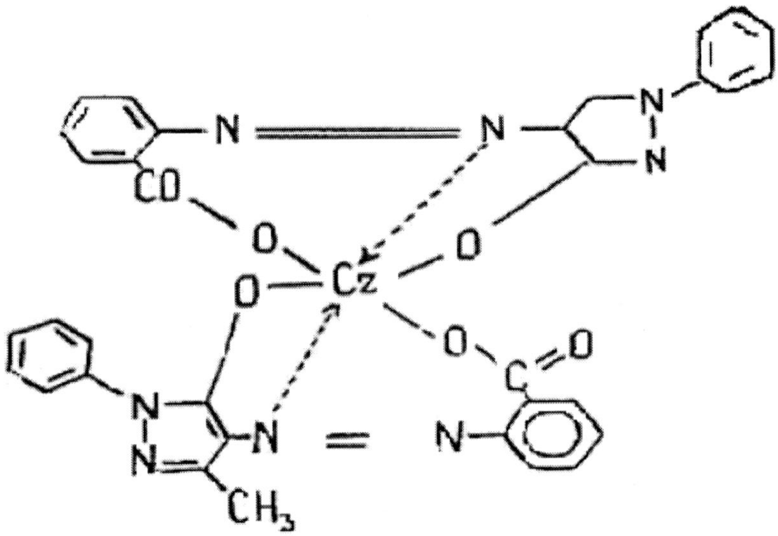

Interesting colours for mass dyeing of synthetic linear PA are obtained by using 1:2 chromium complexes of 0,01 – dioximonoazo-dyes of benzolazo – 1-phenyl – 3-methylpyrasolon and/or benzolazonaphthalene series [98] and chromo – ferrous complex [99].

A great number of patents deal with the derivative of antrachinone [100]. Thermostable and lightfast dye of green colour has been patented [101]. The firm CIBA suggests brown pigments thermostable in PA melt, in the molecule of which two substituted antrachinone residues are condensed by aromatic, more often polycyclic radical of naphthalene type, antrachinone and so on [102, 103].

However, it should be noted that despite the variety of synthesized dyes, both metallocomplex and antrachinone, their use for direct introduction into polymer is limited because of the impurities of inorganic salts.

Dyes of similar type are developed in UIS states [104]. They are produced in the form of salts with amines suitable both for mass dyeing, by introducing into finished polymer, and for dyeing from water baths. Under the name of "caprosols" they appeared much earlier than similar foreign dyes.

1.1.6. Derivatives of Aroilenbenzimidazoles Dyes for Polyamides Dyeing in Mass

Analysis of literature data allows to select one group of dyes synthesis of which is specially intended for mass dyeing of synthetic fibres. There is a variety of dyes – aroilenbenzimidazoles, products of condensation of ortho – (peri-) di – and tetracarboxylic acids with monoamines or ortho- and peri – diamines [104, 105] which belong to this group. As it is known the most valuable are high – temperature dyes of bis-alkyl-imides type of perylene – tetra–carboxylic acid [106], dyeing PA fibres into red, for example: pigment red I, formed at interaction between dianhydride perylene – 3, 4, 9, 10 – tetracarboxylic acid and m – xylidine, of the following structure:

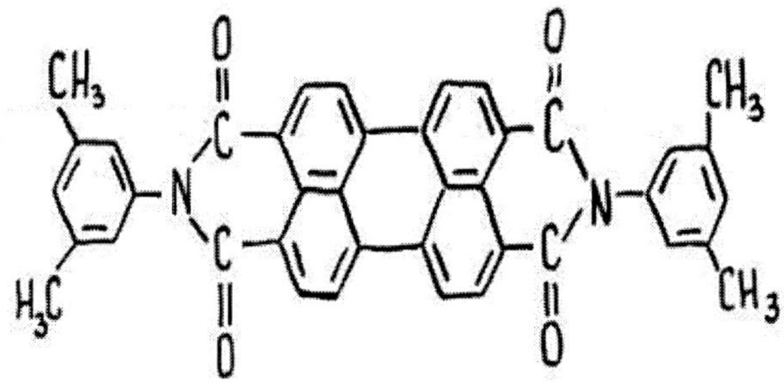

possessing high light fastness. However, it hardly dissolves in polymer melt and requires thorough dispersion.

Interesting works on the synthesis of dyes-derivatives of aroilenbenzimidazoles were done first by the firm "Francolor" and them continued by different researchers [107]. The use of o – phenylene diamine; perinaphthi – lendiamine; 5, 6 –diaminoacetonaphthene; 9, 10 – diaminophenanthrene and 1, 2, 4, 5, - tetraaminobenzole are described as amino component, and the use of the following acids: phthalic, perylene – 3, 4, 9, 10 – tetracarboxylic, piro-mellate, antrachinone - 1, 2 and 2, 3 - dicarboxylic, benzo – 1, 2 – antrachinone – peri – dicarboxylic and also 1, 3 – dicarboxylic aliphatic acids – as the second component. Both components may contain substituents (excepting sulpho group). Universal dye of brown colour may be the example:

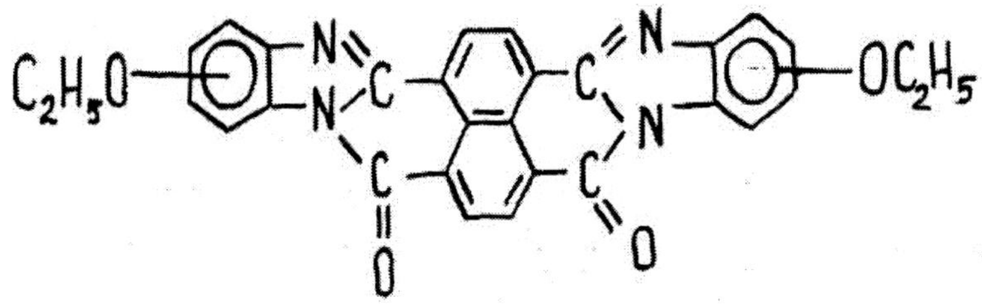

Naphthoilene – bis – benzimidazole and its substituents are of the most interest. Behavior of these dyes in PCA melt for determination of the links between chemical composition of dyes and their thermal stability was investigated in the work [108]. However, in this work they have been unable to determine the effect of substituents in dye molecule on its thermal stability. The cause of high thermal stability of naphthoilene – bis – benzimidazole derivative in PCA melt, having one ethoxy-group (caprosol brown 4K) in each benzole nucleus is not quite clear to the end.

These dyes may be used for mass dyeing of synthetic polymers at synthesis and at direct introduction into melt. Possessing solubility and high stability in PA melt, they, at the same time, have low light fastness of paints.

Compounds of copper with 2 – mercapto-benzimidazole [109], 2 – methylbenzimidazole [110], and oxyphenylbenzimidazole [111, 112], being introduced at any stage of PA-6 polymerization, being, ay the same time, both dyes and light stabilizers have been of great interest lately. Use of complex compounds is usually forced measure because of the difficulties, connected with introduction of metal copper and its salts into PA.

Chromophore system of aroilenbenzimidazole dyes is characterized by the presence of imidazole ring, condensed with aromatic ring and periarilene residue. Forming single conjugated system imidazole ring is conjugated both with benzol and naphthalene residue.

Colouring substances on the basis of naphthoilentetracarboxylic acid for mass dyeing of PA-fibres using polymeric concentrate of the dye [113] were obtained in Rubezhansk branch of SRAPIK. It should be noted that the most part of the dyes of this class are yellow and apply yellow paint on PA fibre and are fully soluble in polymer melt. Dyes of the following structures were tested:

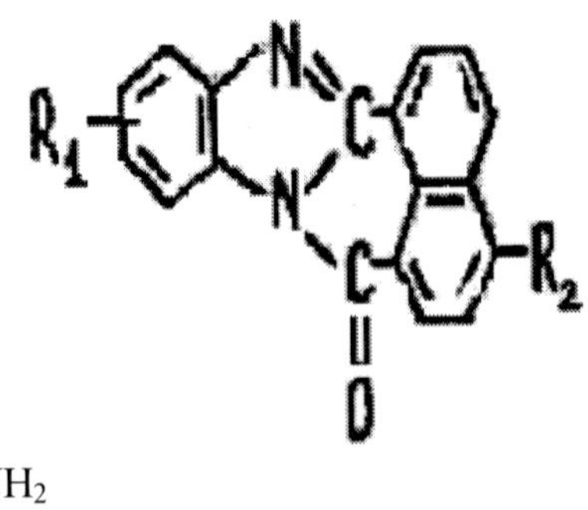

where R_1=H_2Cl R_2=COOH_7NH_2

All the dyess, except those containing R_1=H and R_2=H_2 in the process of obtaining polymeric concentrate of the dye in one way or another reduced concentration, and being dissolved in H_2SO_4 gave insoluble residues, probably the products of interaction with PCA melt. The process of fibres shaping, containing 0,5-1% of the dye, welt on continuosly and there were no difficulties at extracting. Lightfastness was in the range of 3-6 points.

Some of them were recommended to be used in composition with black and caprosol yellow 4K at mass dyeing of PCA through the stage of polymeric concentration of the dye.

Vat aroilenbenzimidazole dyes are of great interest. That the dye might be reduced into leuco compound its molecule must have no less than two conjugated carbonyl groups. For example: 1, 2 – (naph-thoilene – 1, 8) – 4,5 – benzimidazole (I), containing one carbonyl group, cannot be reduced into leuco compound, but mixture of trans – and cys-isomeres of naphthoilene - bis – benzimidazole (II, III)under the name of Vat red 2G easily produces “vat” of olive-green colour.

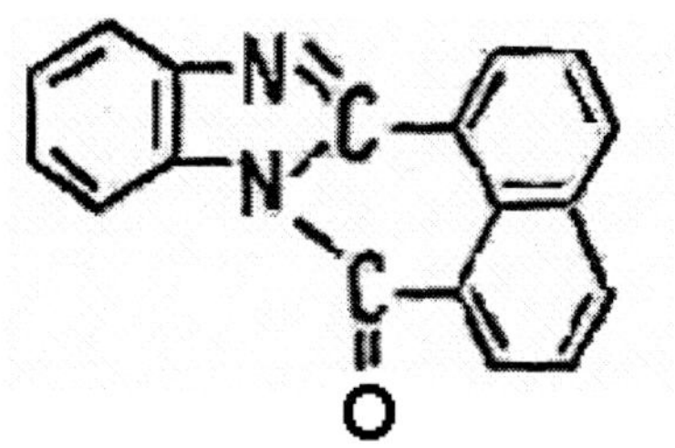

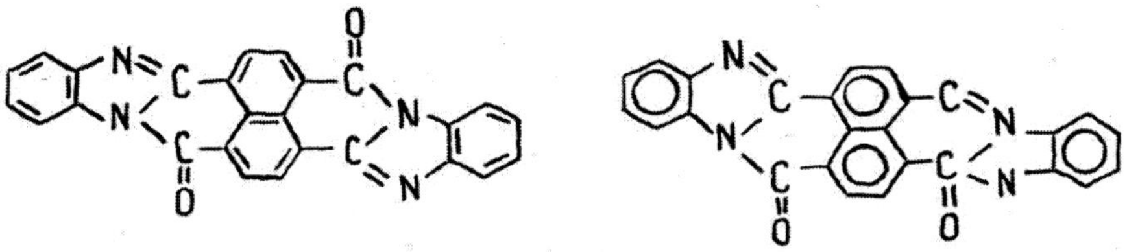

In most cases the ability to be reduced in agressive medium of PA melt is the negative factor, since reduced form of the dye, as a rule, has another colour than initial keto-form, and is characterized by decreased light fastness.

Unfortunately, today there are no clear pre-conditions for dyes selection and in each case this problem may be solved only by empirical way. Nevertheless, the problem of obtaining new thermostable dyes is still urgent.

Daylight – fluorescent pigments, which, in most cases, are solid solutions of luminescent dyes in different polymers are used more often in different fields of science and engineering. Here bright colours are obtained at small concentration of dyes. Some substituted of aroilenbenzimidazole possess fluorescent properties and are used for preparation of daylight – fluorescent dyes [114] and interest in this class of compounds [115] has grown when the opportunity of their use for mass dyeing of polymers has been shown.

Over the last years Kavaliev and his research workers [116, 117] succeeded in synthesizing luminophores in the series of naphthoilenbenzimidazoles, containing fluorophoric groups. For mass dyeing of polymer materials there were proposed aryloxazoline – and phylenoxadiazolil – naphthoilenbenzimidazoles [118, 119]:

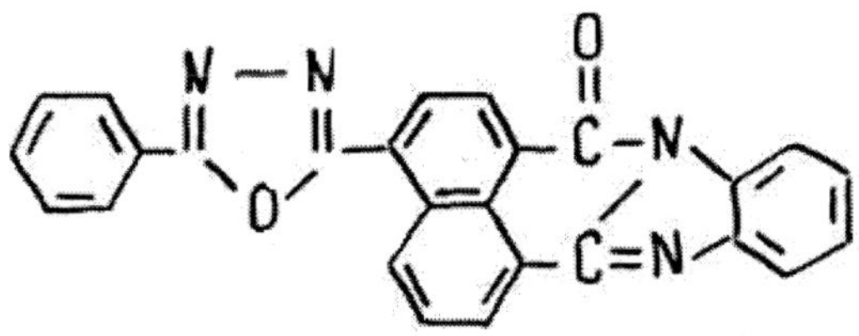

where X=CHN

and also benzoxasolil –, benzimidazolye – and benziazolil – naphthoilenbenzimidazoles – luminophores of yellow glow [120, 121].

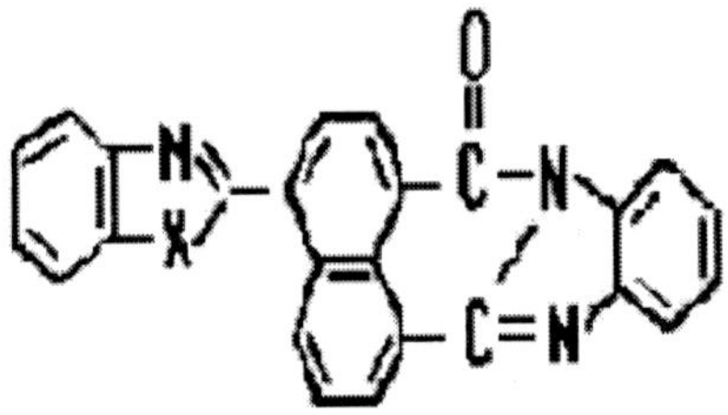

where X=NH, O, S

Further Krasovitskiy and his research workers [122] obtained a number of derivatives of naphthoilenbenzimidazoles both for structurally – dyed polymers and for mass dyeing.

Unfortunately, because of low thermal stability a great number of these dyes found their application for dyeing polyolefines and polystyrenes and in the case of bath dyeing for PA. But despite this interest in them, and first of all from the point of view of practical use of luminophors, does not fall. And of course their relative accessibility plays unimportant role.

Aroilenbenzimidazoles are of great interest, since they are not only compounds modelling structure of "staircase" polymers – polyaroilenbenzimidazoles [123] at different stages of their synthesis, but potential monomers for synthesis of different thermostable polymers, containing phthalimide side-groups [124], copolyaroilenbenzimidazoles of asymmetric structure [125], and also may be dyes for fibre-forming polymers [126, 127], improving their light- and thermal stability. Proceeding from this it was necessary to investigate spectra of luminescence and thermal properties and also their compatibility with PCA melt. Compounds, presented in Table 1, were synthesized on the basis of two-nucleus bis-(0 – nitroamines):3,3 – dinitrobenzidine, 3,3 – dinitro – 4, 4 – diamino – diphenyloxide, 3,3 – dinitro – 4,4 – diamino - diphenylmethane. Transformation of bis(0 – dinitroanilines) intro bis (3 – amino – 4 – phthalimido) – arylenes and definite bis – (1,8 – benzollene – 1,2 - benzimidazoles – 1,2 – benzimidazoles - was carried out according to the following scheme [127]:

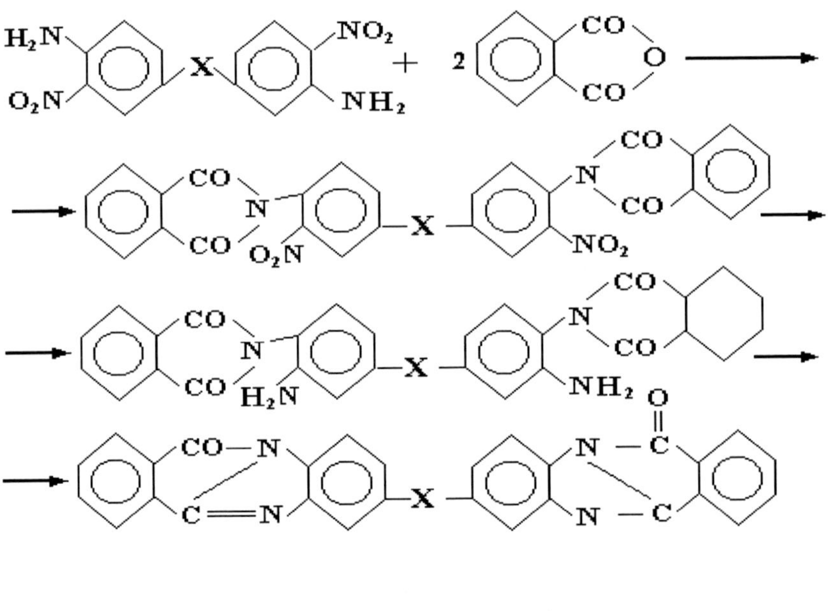

Table 1. Structural forms and characteristics of the derivatives of bis – aroilenbenzimidazole

Number	Compound	T of melting ^0C	T of decomposition ^0C	Molecule weight	Colour
1.	XL	385	356	438	yellow
2.	XLI	317	335	454	lemon - coloured
3.	XLII	315	327	452	light yellow
4.	XLIII	307	386	670	yellow
5.	XLIV	435	388	538	brightly yellow
6.	XLV	335	395	554	lemon – coloured
7.	XLVI	295	370	552	light yellow

Biss – (1,8 – naphthoilene – 1,2 – benzimidazole) was synthesized at interaction of bis – (0 – phenylene diamines) of different structure with naphthalic anhydride according to the scheme [127]:

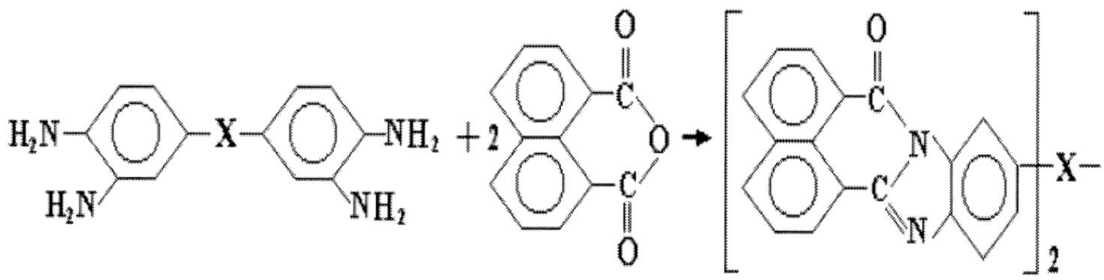

Electronic spectra of obtained compounds and presented in Figures 1.1 and 1.2 are characterized by absorption at λ_{max}^{em}=410-430 nm and luminescence at λ_{max}^{em}=510=530 nm . Clear peaks of spectra of synthesized compounds luminescence speak about high purity of obtained products.

Criteria of compounds thermostability were temperatures: of melting, beginning of compounds decomposition, loss of initial mass by 5, 10 and 25%. These data are given in Table 2.

Table 2. Thermal characteristics of bis – aroilenbenzimidazoles

Compounds	T of melting, ^0C	T of decomposition beginning, ^0C	T of mass loss ^0C		
			5%	10%	25%
XL	385	356	454	495	562
XLI	317	335	415	473	528
XLII	315	327	409	438	507
XLIII	307	386	442	463	480
XLIV	435	388	497	545	637
XLV	335	395	440	517	563
XLVI	295	370	473	508	605

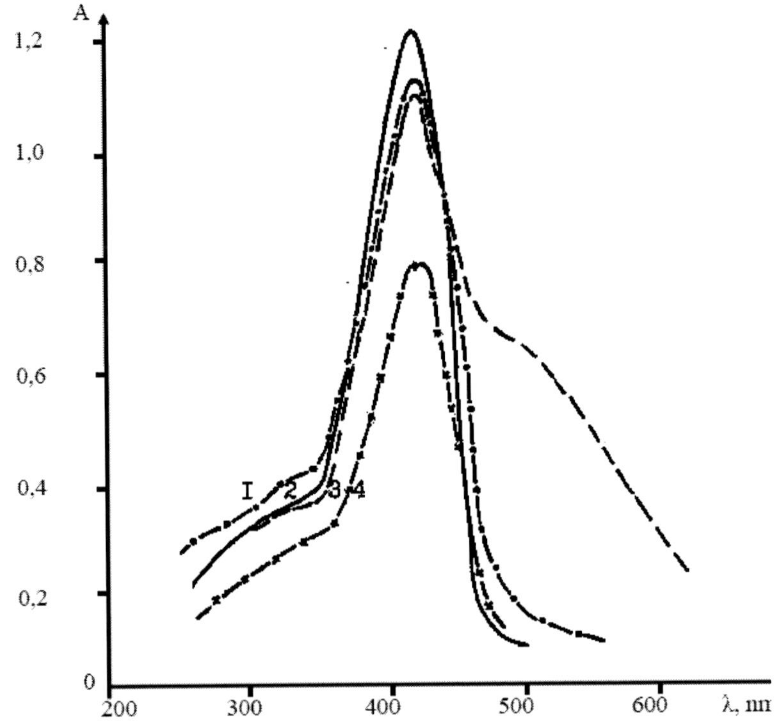

Figure 1.1. Absorption spectra of derivatives of bis – aroilenbenzimidazole: XLV – (1), XLVI (2), XLVII (3), XLIV- (4).

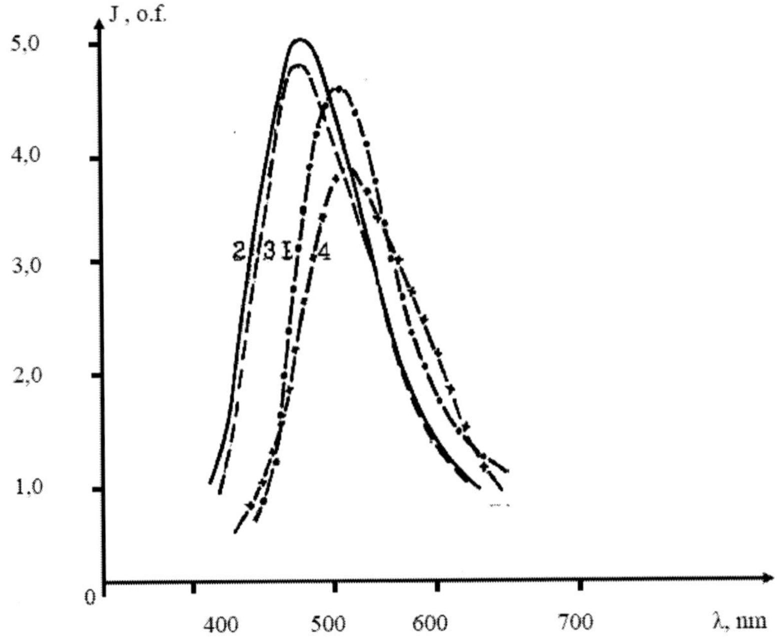

Figure 1.2. Luminescence spectra of derivatives of bis – aroilenbenzimidazole at excited λ = 365 nm: XLV – (1), XLIV – (2), XLIII – (3), XLVI – (4).

As it is seen from Table 2.2 all compounds, being investigated, possess high thermal stability. Temperatures of melting and beginning of decomposition are within the range T_{melt}=295-435 ^{0}C and T_{dec}=327-395 ^{0}C and this shows potential possibility to introduce compounds XL-XLIII into PCA melt, as their thermal stability exceeds the temperature of PCA production and shaping.

Investigations on solidity and stability of XL-XLVI compounds at the stage of ε-caprolactam polymerization were carried out to determine compatibility and stability of bis – (1', 8' – benzoilene – 1,2 – benzimidazoles) and bis – (1', 8' - naphthoilene – 1,2 – benzimidazoles) in reactive PCA mass. Obtained coloured polymer was investigated visually to homogeneity of solid solution, its solubility in H_2SO_4, in the mixture of 80% formic and trifluoro – acetic acids in ratio 1:1 and in m – cresol. Ability of PCA melt to form into a fibre, and also molecular masses of dyed and undyed polymer with average viscosity were defined too.

Polymerization of ε – caprolactam in the presence of XL – XLVI compounds was carried out in ampoules. Thick transparent polymer mass dyed in different colours (from lemon colour to yellow) depending on applied compound was obtained in all ampoules after 6-hours' synthesis. When using all the compounds initial colour of the melt did not change. This shows stability of investigated compounds in polymer melt.

Effect of dyes on moldability of PCA melt was defined by introducing polymer mixture into molding machine where polymer was melted in nitrogen atmosphere, homogenized with the dye and then at passing of the melt through spinneret, provided with metal sieves, formed into PCA monofilament.

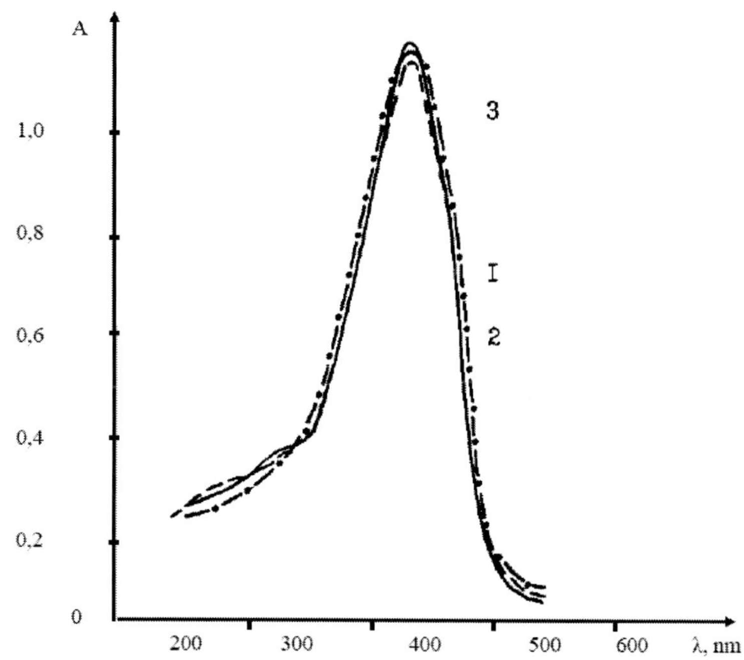

Figure 1.3. Absorption spectra of solutions of the dye XLV – (I) and dyed PCA before the spinneret (2) and after spinneret (3) in the mixture of formic and trifluoro – acetic acids (1:1).

Photographs of dyed and undyed fibres and their cross-sections at 20-times increase, where homogeneity of polymer structure is distinct, show good solubility of bis-aroilenbenzimidazole derivatives in PCA melt. It should be noted that the process of fibres formation, containing XL-XLVI dyes, went on steadily and there were no difficulties during extraction.

Figure 1.3 present spectra of XLV dye solution absorbtion in the mixture of formic and trifluoro-acetic acids, solutions of dyed fibres before and after spinneret, which are practically laid on each other, showing full solution of the dye and good homogeneization of polymer melt. It should be noted that the samples of fibres, being studied, fully dissolved in solvents without formation of residues (possible products of dye and PCA interaction), which tells about inertia of components regarding each other. Blinding screen was caused mainly by impurities of mechanical nature.

Besides, tests on possibility of XL-XLYI compounds use for mass dyeing of PCA through the stage of obtaining polymeric concentrate of the dye (PCA).

In the process of polyamidation foaming of the polymer was not observed. Thermal stability of the dyes is satisfactory at warming – up for 48 hours and this shows that their introduction at the stage of ε – caprolactam polyamidation does not deteriorate PCA properties.

Data, given in Table 3, show that small amounts of the dye (up to 2% of mass) do not exert essential effect on molecular weight of polymer. It should be noted that polymer gains bright colour even at dye concentration of 1% of mass.

Table 3. Molecular weight of synthesized PCA, dyed by XLY compound through the stage of polymeric concentration of the dye

Number	Polymer sample	Additive concentration	η	Molecular weight
1.	Initial PCD	0	0,61	18000
2.	PCD+XLY dye	0,25	0,62	18600
3.	-//-	0,50	0,61	18000
4.	-//-	1,00	0,59	17500
5.	-//-	2,00	0,60	17800
6.	-//-	5,00	0,45	12000

Thus, all compounds, being tested, dissolve and combine with PCA melt quite well and do not influence the polymer negatively. On the other hand, these dyes are stable in polymer melt.

1.2. Photooxidative Destruction of Dyed PCA

Breaks of the main chain of macromolecules are the main reasons for the loss of PCA service properties at photooxidative destruction; therefore definition of the change of polymer molecular weight before and after photooxidation is one of the methods of evaluating such important characteristic as the number of breaks. There were investigated PCA samples obtained during introducing 0,25-0,5 mass % of dye into reactive mass both at the stage of polymerization and directly into polymer melt.

Table 4 presents data on conservation of molecular mass of the crumb of dyed and undyed PCA samples before and after irradiation, and this shows great increase of photooxidative stability of dyed PCA samples.

Table 4. Conservation of molecular mass of synthesized PCA stabilized by XLY compound after 10-hours irradiation by mercury-quartz lamp

Additive concentration %	[ή]		Molecular mass		Conservation of molecular mass, %	A number of break
	Before irradiation	After irradiation	Before irradiation	After irradiation		
-	0,66	0,44	21000	12000	59,5	0,76
0,25	0,62	0,51	18600	14500	77,9	0,28
0,50	0,61	0,53	18200	15200	83,5	0,20
1,00	0,59	0,58	17500	17100	97,8	0,02
2,00	0,60	0,57	17800	16700	93,7	0,07
5,00	0,45	0,41	12300	10900	88,7	0,13

It should be noted that in the process of dissolving irradiated polymer in H_2SO_4 and m – cresol formation of insoluble products was not found, though data [128] obtained during irradiation of PCA film in vacuum by complete spectrum of the lamp PRK-2 at 30^0C for 10 hours showed the decrease of PCA solubility by 13% in m-cresol.

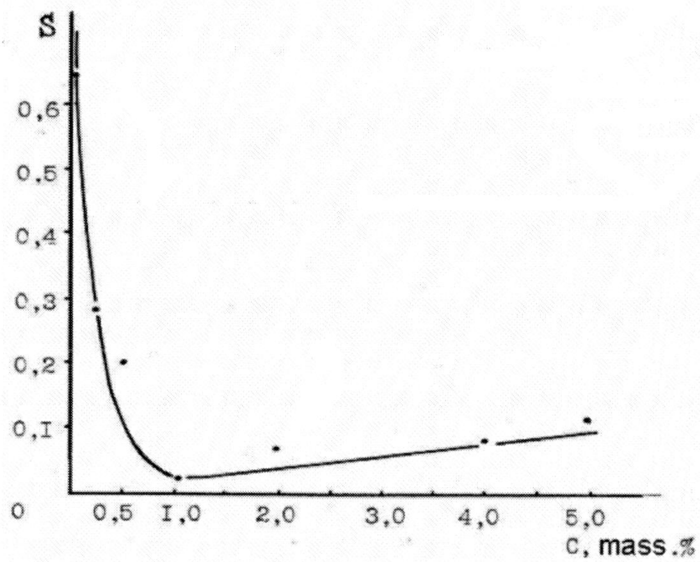

Figure 1.4. Dependence of the number of breaks (3) of PCA macromolecules on additive concentration.

As it was mentioned earlier, diffusive restruction of oxygen penetration into polymer matrix might play here a great role. That is why diffusive factor may be ignored at small size of particles of polymer crumb or small thickness of obtained PCA fibres.

From the data of Table 4 it follows that stabilizing action of XLY compound depends on its concentration in polymer, besides when concentration increases up to 1 mass % conservation of molecular mass rises, but at further growth of the concentration it decreases a little.

All this is displayd in Figure 1.4, where concentration dependence of the number of breaks of PCA macromolecules for XLY compound is presented.

Such shape of a curve may be explained by the fact that at increase of additive content first, light absorption increases (as content of chromophore groups increases), secondly, supermolecular structure of polymer itself changes. Similar investigations were carried out and with dyed PCA fibres, obtained during additive introduction into polymer melt.

Table 5. Conservation of molecular mass of PCA fibres stabilized by bis – aroilenbenzimidazoles (1,0 mass %) after 24-hours irradiation by mercury-quartz lamp

Additive concentration %	[ή]		Molecular mass		Conservation of molecular mass, %	A number of break
	Before irradiation	After irradiation	Before irradiation	After irradiation		
PCA initial	1,02	0,53	35200	15000	42,7	1,34
PCA + additive XL	1,04	0,95	35900	32300	89,8	0,11
-//- XLI	0,97	0,92	32800	30700	93,4	0,07
-//- XLII	1,00	0,89	34300	29400	85,6	0,17
-//- XLIII	1,10	1,00	38800	34400	88,6	0,13
-//- XLIV	1,07	1,00	37600	34500	91,7	0,09
-//- XLV	1,02	0,98	35400	33400	94,3	0,06
-//- XLVI	1,05	0,97	36700	32800	89,4	0,12

As it is seen from the data of Table 5, initial polymer loses more than 40% of initial molecular mass after 24 hours irradiation. All additives have stabilizing effect, expressed in conservation of molecular mass from 85,6 up to 93,4% depending on dye structure.

Comparing similar by "bridge" groups compounds of benzoilenbenzimidazole (XL – XLII) and naphthoilenbenzimidazole (XLIY – XLY) series we note that bridge group, but not chromophore part, exerts great effect on stabilizing action. This is proved by close value of indexes of conservation of molecular mass and breaking numbers accordingly for compounds XL-XLIY, XLI-XLY, XLII-XLYI. Compounds with oxygen bridge have more intensive light protective effect. Compounds with oxygen bridge possess greater mobility around 0 – atom, then benzidine derivatives XL-XLIY, and at the same time greater degree of conjugation, then compounds with methylene bridge XLII-XLYI.

1.3. Physico-Mechanical Characteristics of Stabilized PCA Fibres at Photooxidation

PCA fibres, in one way or another, bear mechanical load in real service conditions. That is why, strength indexes and, in the first place, rupture stress and extension are of great practical importance; change of these indexes at fibres irradiation may characterize photooxidative stability of the polymer.

Fibres, containing derivatives of bis-aroilenbenzimidazole, were subjected to ultra-violet light action of the lamp PRK-2. Mechanical properties of the fibres were determined depending on cross section area. Value of rupture stress (σ) is the mean value of true rupture stress obtained from indexes of tearing machine.

Stabilized fibres, containing 0,6 mass % of the XLY compound, were obtained by introducing additive at the stage of producing polymeric concentrate of the dye and the others – by usual method of polymer crumb powdering before polymer melting and fibre forming. Data on physico – mechanical indexes of PCA fibres stabilized by XLY compounds are given in Table 6. From these data it follows that stabilized fibres greatly exceed unstabilized ones according to photooxidative stability.

Table 6. Physico-mechanical indexes of stabilized polycaproamide fibre before and after 24-hours irradiation by mercury-quartz lamp

Polymer sample	Concentration of additive, %	Rupture stress σ, Mpa		Rupture extension, ε, %		Conservation, %	
		before irradiation	after irradiation	before irradiation	after irradiation	before irradiation	after irradiation
PCA initial	-	325,68	154,99	122	86	47,6	70,5
PCA + XLY additive	0,25	333,48	264,01	128	90	79,2	70,3
-//-	0,50	340,76	285,69	130	98	83,8	75,4
-//-	0,60*	339,35	284,51	124	97	83,9	78,2
-//-	1,00	341,73	317,12	116	106	92,8	91,4
-//-	2,00	335,30	277,62	118	91	82,8	77,1

*was obtained through the stage of polymeric concentrate of the dye.

The reason of increased photooxidative stability is not only stabilizing activity of bis-aroilenbenzimidazoles derivatives but also the effect of these compounds on supermolecular structure of polymer matrix. As it is seen from the Table 6 physico – mechanical indexes of polymer do not practically depend on the method of additives introduction. Data of Table 6. agree with data of Table 5. In all cases the greatest degree of conservation of polymer initial properties is registered at the concentration of XLY compound – 1 mass %.

1.4. Effect of Derivatives of Bis-Aroilenbenzimidazole on Supermolecular Structure of PCA – Fibres

Electron – microscopic photographs of dyed and undyed PCA fibres before and after ultra-violet irradiation were obtained to clarify possible effect of bis-aroilenbenzimidazole derivatives on supermolecular structure of PCA fibres by the method of raster electronic microscopy (REM) with increase of 12000 times. Any specific details of the structure were not observed in the initial sample. After irradiation the initial sample undergoes great changes. Its surface is completely covered by the particles of irregular shape: hollows, holes, cracks (products of decay) which show the destruction of surface layer in the process of photooxidation.

Stabilized sample does not practically undergo great changes after irradiation. Its surface is smooth without visible traces of products of decay. It is seen here that presence of stabilizing compounds in polymer macromolecule inhibits the process of photodestruction.

It should be also noted that modified unirradiated fibre displays more distinct tendency to crystallization. There were found plate-like and even spherulitic formations. In our opinion this is connected with the fact that derivatives of bis-aroilenbenzimidazole play the role of peculiar centres of crystallization and this causes change of molecular orientaton in the fibre at the moment of forming when supermolecular structure begins to organize. Besides, new more regulated structure, typical for ordered state, is formed which agrees well with conclusions of the works [129, 130]. The authors consider that dyes influence the rate of formation of nucleation centres at the expence of decreasing surface energy of crystallites leading to greater degree of order of the structure and decrease of crystallites critical sizes.

Structural changes in dyed fibres improve stability of PCA fibre to photooxidative destruction. Data of electron microscopy are proved by X-ray study of initial and stabilized fibres before and after irradiation.

X – ray studies of initial and dyed by bis – (1', 8' – naphthoilene – 1, 2 - benzimidazole) oxide (XLY) PCA studies have been carried out with the purpose of investigating possible changes in amorphous – crystalline structure of PCA fibre at the time of introducing such volumetric dyes as bis-aroilenbenzimidazoles and their distribution in the fibre.

Data, obtained during the analysis of large-angle ionization X-rayograms, characterizing the structure of both dyed and undyed PCA-fibres are given in Table 7.

Identity of the angles of diffraction maximums and equality of their hemispheres show that introduction of XLY does not change interplane distances of polymer crystallites and their sizes.

The degree of polymer crystallinity, which characterizes the share of regularly packed molecules is estimated by the intensity of dispersion in maximum 1 m. Some differences connected with introduction of a dye are observed, namely intensity of dispersion is a bit lower both along the equator and meridian. This speaks about the fact that introduction of the additive leads to the decrease of a number of reflective planes (decrease of reflection centres), which, in its turn, is connected either with deterioration of polymer crystalline structure or with rotation of reflective planes.

Results of small-angle meridional measurements (Figure1.5.) show the absence of small-angle diffraction maximum and this shows the absence of large-period structure, that is in this case there is no regular alternation of crystalline and amorphous sections along the axis of the fibre.

Presence of small-angle diffusion dispersion speaks about the presence of submicrovoids (SMV) in the fibre. Rearranging SADD according to the data of Figure1.5 in coordinates 1g I from φ^2 (Figure 1.6.) by the procedure described in the work [131] there may be defined the size of H_n ($H_n = 1,29\sqrt{\frac{\Delta \lg I}{\Delta\varphi}}$) and concentration of SMV – (N_{tr})(Assuming the form of SMV as spherical).

Table 7. Data of X-ray diffraction analysis of dyed and undyed PCA-fibre

Structural characteristics	PCA-fibre	
	dyed	undyed
Intensity of diffraction maximum 1 m of relative unit	40	59
Angle of dispersion 2^{0}	$10^{0}30$	$10^{0}40$
Average angle of crystallites disorientation $\Delta_{\varphi \ meas}$	150	130
Degree of crystallinity C_K, %	61	85
Size of crystallites:		
longitudinal, A°	85	102
transversal, A°	50	52

Sizes of both large and small SMV and their concentration have been defined in connection with the fact that dependence of lg on φ^2 is not linear. Results of small-angle X-rays measurements, which show that sizes of large SMV are 260-290 A and small- 60-80 A with concentration of the order of 10^{14} and 10^{15} cm^{-3}, accordingly, are given in Table 8.

As it is seen from Table 8 the sizes of large SMV and, hence, their volume is decreasing during fibres dyeing. It speaks about the fact that molecules of the dye enter into pores of SMV. The dye enters into small SMV and besides the concentration of small SMV ($N_{O\partial}$) decreases by 20 times for factory fibres and 5 times – for laboratory one.

Table 8. Results of small-angle X-ray measurements of PCA fibres

PCA-fibre		Large SMV			Small SMV		
		Sizes of voids, A	Volume of voids, $cm^3 \times 10^{17}$	Concentration of voids, $cm^3 \times 10^{14}$	Sizes of voids, A	Volume of voids, $cm^{-2}x10^{19}$	Concentration of voids, $cm^{-3}x10^{15}$
Factory	Initial	282	1,18	3,14	61	1,24	47,70
	Dyed	264	0,96	3,5	69	1,73	2,23
Laboratory	Initial	288	1,25	2,26	79	2,58	11,30
	Dyed	260	0,93	3,38	69	1,73	2,23

Thus, the dye enters into small SMV filling in space of pores and, in addition, facilitates fibre hardening, which, in its turn, influences photooxidative stability of PCA-fibres. These data are proved by the results of physico-mechanical researches showing that dyed fibres greatly exceed initial ones in photooxidative stability (Table 6).

Dye molecules, filling in SMV space, are connected with polymer molecules by forces of physico-chemical interaction, at the expence of which load on cross-section of the fibre is distributed more uniformly during the loading of dyed fibre and this leads to hardening of fibres.

Thus, it is seen that thermostable dye XLY bis-(naphthoilenbenzimidazole) oxide facilitates hardenening of PCA-fibres at ultra-violet irradiation when it is introduced into PCA mass.

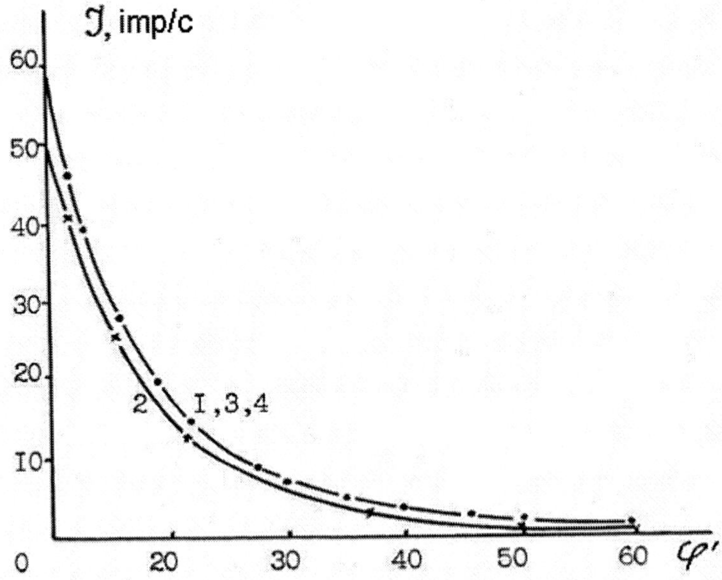

Figure 1.5. Curves of small-angle meridional measurements of PCA-fibres: initial of laboratory (2) and factory shaping (I), stabilized XLY of laboratory (3) and factory (4) shaping.

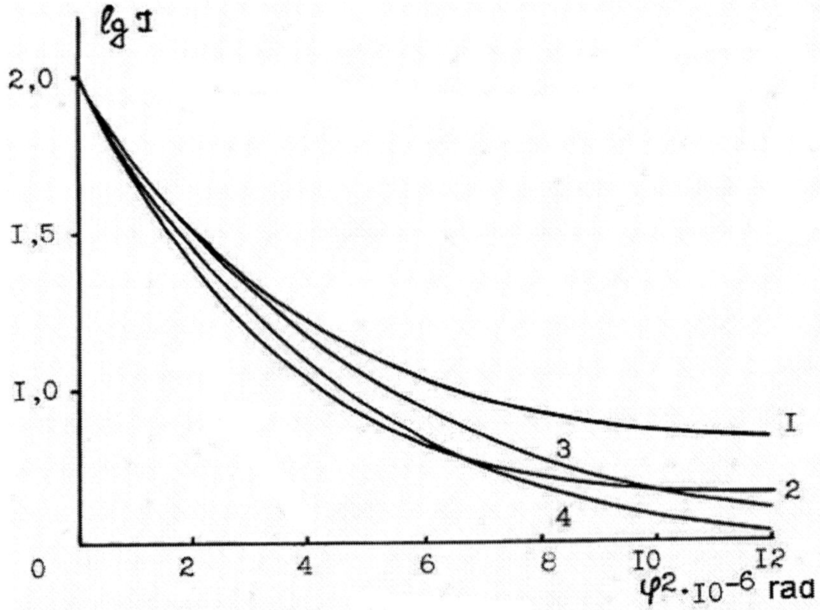

Figure 1.6. Logarithmic dependence of small-angle meridional measurements of PCA-fibres: initial of laboratory (2) and factory (1) shaping, stabilized XLY of laboratory (3) and factory (4) shaping.

1.5. Mechanism of Light-Stabilizing Action of Bis-Aroilenbenzimidazoles

Breaks of macromolecule main chain, being the results of complex set of processes taking place in polymer, following primary acts of light and oxygen absorption, cannot characterize the process of oxidative destruction on the whole, measure of which may be only oxygen absorption [132]. Method of determining oxygen absorption during photooxidation of polymers is highly sensitive and allows to estimate the efficiency of stabilizers action at early stages of photooxidation at small polymer weights, at low intensities of irradiation, which gives the possibility to create conditions of artificial ageing, approximate to natural.

1.5.1. Absorption of Oxygen by Polycaproamide Stabilized by Aroilenbenzimidazole Derivatives

Taking into account the most important role of oxidative processes at PCA light ageing (Section 1.1) it was reasonable to investigate oxygen absorption by initial and stabilized PCA-fibre during long-wave irradiation by filtered light of the lamp DRSH-1000, because under the action of long-wave light, which is not absorbed by chromophore amide group of the polymer, photooxidation of PCA is defined by initiation action of impurities, products of oxidation and also special additives (pigments, dyes and so on) [16].

Curves of the dependence of absorbed oxygen amount per unit of polymer mass on the irradiation period at the given temperature are given in Figures 1.7-1.9. Analysis of kinetic curves of oxygen absorption shows that at initial stage photooxidation is going on at a great speed but after some time this process slows down and at last goes into stationary mode.

Probably at the initial stage of irradiation there takes place oxidation of impurities or by-products being formed in the polymer during the process of its production and processing. It may be supposed that formation and determination of stationary concentration of ketoimide product [26] responsible for the process of PCA oxidative destruction plays a great role. It should be noted that in our experiments stationary mode is set up for 3-5 hours depending on the nature of irradiated samples. After the light source is turned off the process of oxidation gradually inhibits up to complete stoppage. However, repeated turn on of the light leads to setting stitionary mode up already in 0,5 hour, moreover this pause in irradiation may be rather long – a day or more.

All this shows that photooxidation of impurities in PCA at the initial stage does not influence further course of the process of PCA oxidation. That is why the rate of oxidation in stationary mode may be used for quantitative evaluation of photooxidative stability of PCA fibres.

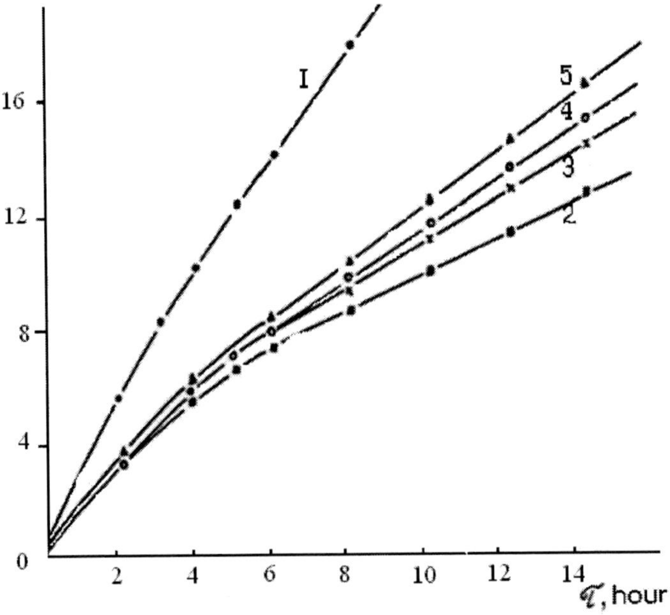

Figure 1.7. Kinetic curves of oxygen absorption by PCA samples: initial (1) and containing XLY in concentration – 0,25 (5); 0,5 (4); 1,0 (3); 2,0 (2) mass % of XLY.

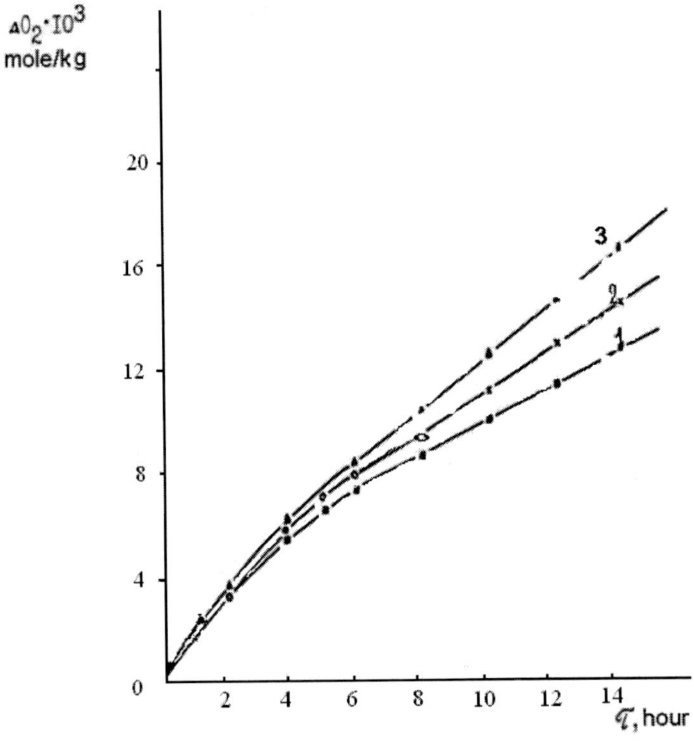

Figure 1.8. Kinetic curves of oxygen absorption by PCA samples, containing derivatives of aroilenbenzimidazole I mass % of XLI-(1), XL-(2), XLII-(3).

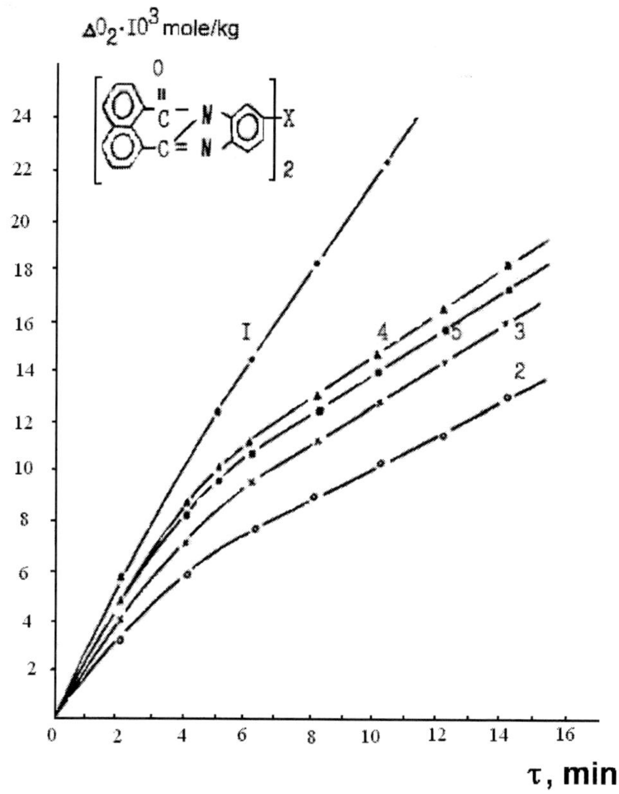

Figure 1.9. Kinetic curves of O 2 absorption by PCA samples: initial – (1), containing I mass % of bis – aroilenbenzimidazole XLY –(2), XLIY-(3), XLIII-(4), XLYI-(5).

Obtained values of the rate of oxygen absorption in the process of PCA photooxidative destruction are given in Table 9.

Table 9. Kinetics of oxygen absorption by dyed and undyed PCA fibres-derivatives of bis-aroilenbenzimidazole

Polymer sample	Additive concentration, mass %	$W_{02} \times 10^{-3}$ Mole/kg • h	i φ	W_0/W_{ST}	W_0/W_{ST} • i φ
PCA – initial	-	2,03	-	-	-
PCA+additive XLIII	1,0	1,00	1,43	2,0	1,40
"-" XLIV	1,0	0,96	1,32	2,1	1,59
"-" XLVI	1,0	0,98	1,48	2,0	1,35
"-" XLV	1,0	0,81	1,53	2,5	1,63
"-"	0,25	1,10	1,18	1,8	1,54
"-"	0,5	0,92	1,25	2,2	1,60
"-"	2,0	0,65	1,70	3,1	1,82

From the Table data it follows that all used compounds in one way or another possess stabilizing activity. They decrease the rate of oxygen absorption by 2-3 times, moreover the greatest stabilizing effect is found, as it has been observed earlier, in the compound XLJ, containing oxygen in a bridge group.

It should be noted that when me increase concentration of the additive, stabilizing effect of bis-aroilenbenzimidazole also increases (Figure1.10). Since compounds XL_XLYI absorb active light contribution of the effect of shielding into the total effect of stabilization of bis-(1',8' –benzoilen – 1,2 – benzimidazoles) and bis- 1',8' – naphthoilen – 1,2 – benzimidazoles) has been estimated.

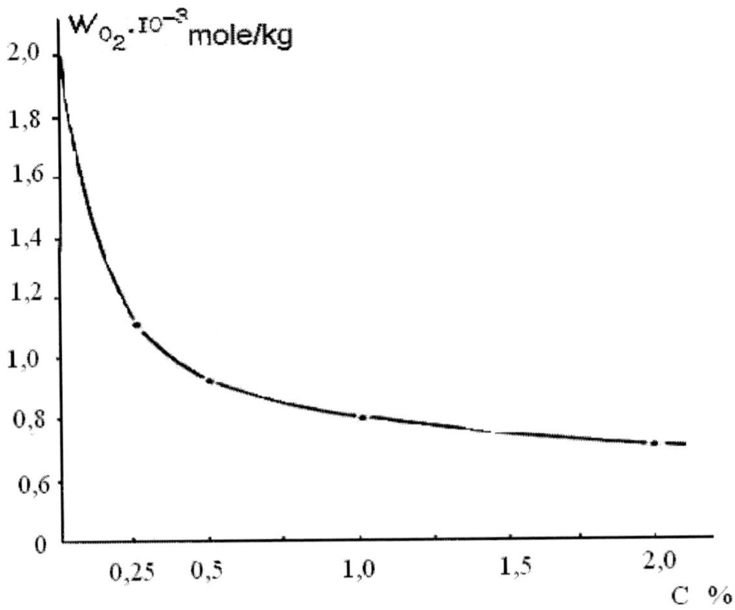

Figure 1. 10. Dependence of the rate of oxygen absorption by PCA fibres, containing compoundl XLY, on the additive concentration.

Values of extinction and shielding coefficients on the wave length of active light (365 nm) were calculated according to the spectra of PCA solutions absorption. The polymer itself absorbs very little in this region.

When comparing coefficients of photooxidation retardation, defined as relations of stationary rates of oxygen absorption by the polymer in the absence of (W_o) and presence (W_{st}) of the additive (Table 9) and coefficients of shielding (iφ), obtained on the basis of spectral characteristics of dyed PCA it is seen that stabilizing effect, achieved only by additives shielding is less than observed stabilization effect, obtained during using bis-aroilenbenzimidazoles. Hence, XL-XLYI being investigated protect PCA not only according to shielding mechanism, as the value of ($W_o/W_{st}\cdot$ iφ) is much more than one.

If protective action of the additive is completely brought to shielding, then $W_o/W_{st}\cdot$ iφ=1. If $W_o/W_{st}\cdot$ iφ <1 – the additive acts as photosensitizer; if $W_o/W_{st}\cdot$ iφ>1 – additional effect of super-shielding is found here.

Other possible mechanisms of PCA light stabilization may be: suppression of PCA excited states and inhibition of radical processes. That is why, possibility of compounds XL-XLYI participation in the process of deactivation of PCA excited states should be considered.

1.6. Spectral – Luminescent Investigations of PCA, Containing Additives of Bis-Aroilenbenzimidazole

1.6.1. Spectral – Luminescent Properties of PCA

Before considering the action of derivatives of bis-(1,8-benzoilen-1,2-benzimidazole) and bis-(1,8-naphthoilen-1,2-benzimidazole), as possible suppressors of polymer exited states, it was necessary to study fluorescence of initial PCA.

Spectra of fluorescence of unirradiated and thermo-unprocessed PCA film are shown in Figure 1.11. These spectra show that maximums of fluorescence wave-lengths (λ_{max}^{em}):375, 425 and 430 nm correspond to different wave-lengths of exciting light (290nm, 350nm, 365nm), moreover intensity of the band 340nm is twice higher than others. All this speaks about the presence of several glow centres in PCA which may be classified as impurities [133] and products of polymer oxidation [20], being formed during polymerization.

Fluorescence spectra of fibres differ from fluorescence spectra of films by small shift (3-5nm) into the region with more long waves. This difference may be explained by distinction in physical structure and degree of macromolecules orientation during samples production. Fibres have greater crystallinity in comparison with the film, which is proved by X-ray diffraction analyses.

Effect of thermal processing of PCA on its luminescent properties has been considered. Thermal processing for 4-8 hours leads to the increase of peaks intensity λ_{max}^{em}=375 and 425nm (by 5-7 times) with simultaneous shift of the band λ_{max}^{em}=375nm up to 385nm into long-wave region and the band λ_{max}^{em}=425nm into short-wave region 410nm, at the time when intensity of the peak λ_{max}^{em}=430nm increases almost by 30 times.

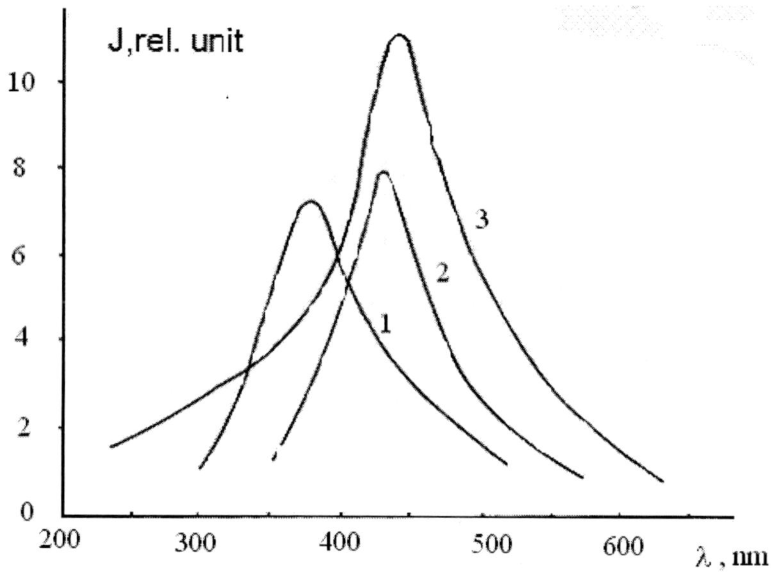

Figure 1.11. Fluorescence spectra of unirradiated and thermo-unprocessed PCA film at the exciting light wave-length of 290nm(1), 350nm(2),365nm(3).

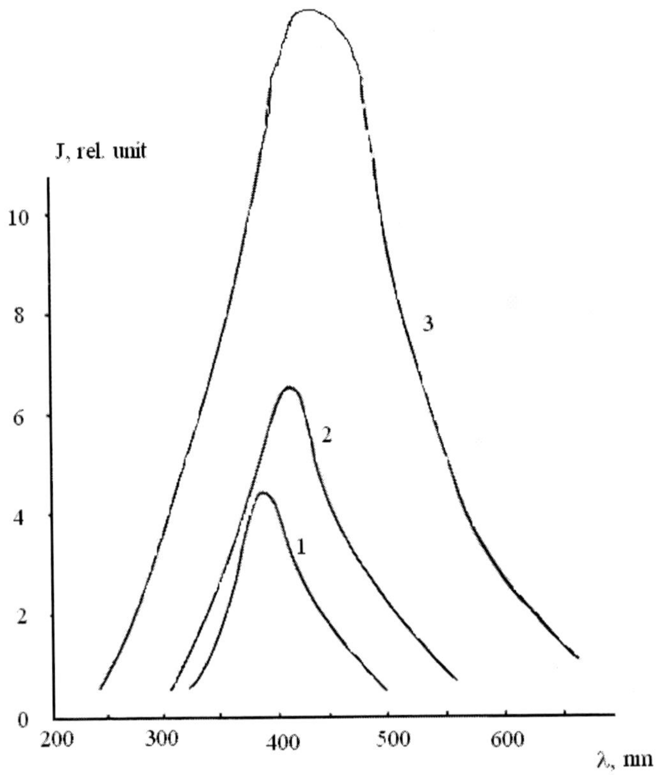

Figure 1.12. Spectra of fluorescence of thermo-processed PCA film during 8-16 hours at the wavelength of exciting light of 290nm(1), 350nm(2),365nm(3).

Further thermal processing leads to gradual decrease of intensity of peakλ_{max}^{em}=385nm and intensity of the band λ_{max}^{em}=430nm after slight decrease (by 1,5 times) becomes constant (Figure 1.12.).

During irradiation of PCA samples by light λ_{max}^{em}=365nm for 2-3 hours increase of peaks intensity is being observed, but in different relations: for λ_{max}^{ex}=290nm – by 7 times, λ_{max}^{ex}=350nm – by 10 times, λ_{max}^{ex}=365nm – by 20-25 times with the shift of fluorescence bands accordingly up to λ_{max}^{em}=385, 408 and 428 nm.

Further irradiation leads to decrease of intensity of the first two peaks and constant value of the peak λ_{max}^{em}=428nm, which shows that the system has gone into some stationary mode. Besides, the intensity of the peak.

λ_{max}^{em}=428 exceeds the intensities of other peaks by many times. So we may come to a conclusion that fluorescence at thermo and photooxidation of PCA is connected with accumulation of one and the same compounds.

Thus, obtained data agree well with literature ones [22, 23], where it is noted that change of PCA glow intensity takes places simultaneously with the change of C=0 groups formed during photooxidation. Authors, classifying PCA glow as formation of compounds of ketomide structure [24, 25] in polymer, come to the same conclusion. These data evoke special interest to PCA fluorescence within the limits of λ_{max}^{em}=428-430nm, excited by the light λ_{max}^{ex}=365nm.

In this connection, investigation of spectral – luminescent properties of PCA, containing additives of bis-aroilenbenzimidazoles was carried out to evaluate the role of suppression mechanism.

1.6.2. Spectral-Luminescent Properties of Stabilized PCA

According to the rule of ultra-violet absorbers selection the most effective stabilizers must absorb not only in the field of absorption of polymer itself, but in the field of its luminescence [24] that is the condition for realization of suppression mechanism of polymer excited state.

Spectra of absorption and fluorescence of dyed compound XLY and initial PCA are given in Figure 1.13. Characteristic peak of initial PCA fluorescence λ_{max}^{em}=425nm (curve I) is absent in dyed polymer (curve 2). At the same time there appears new peak λ_{max}^{em}=530nm corresponding to the emission spectrum of compound XLY (Figure 1.12.).

It should be noted that spectrum of dyed PCA excitation (Figure1.13., curve 3) completely coincides with the spectrum of initial PCA fluorescence.

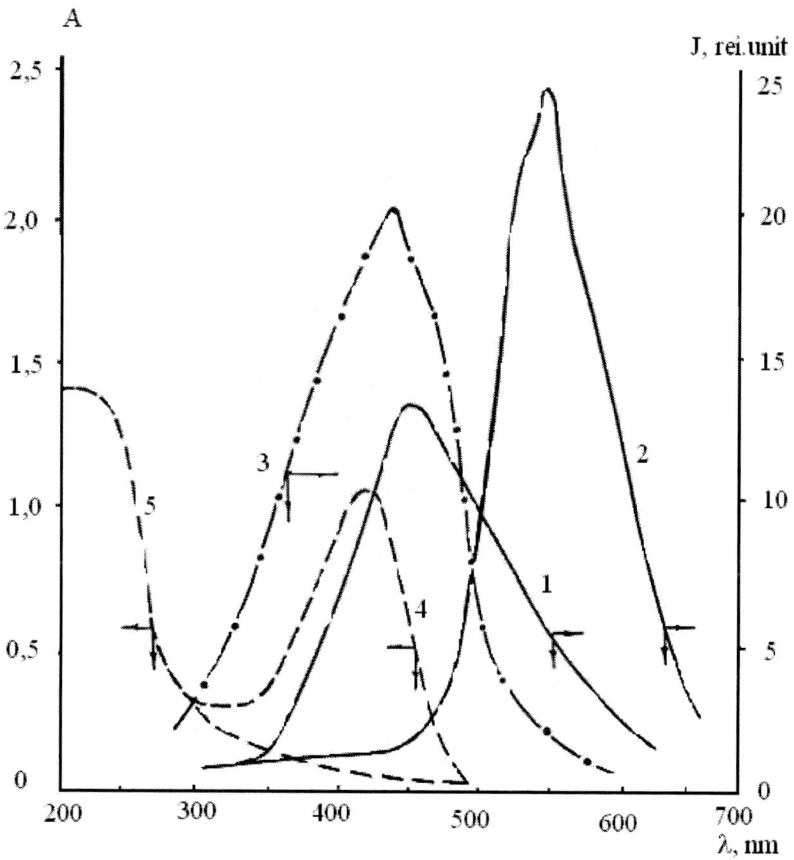

Figure 1.13. Spectra of absorption and luminescence of dyed and undyed PCA fibres: 1) luminescence of initial PCA, 2) luminescence of PCA, dyed by XLY, 3) excitation of PCA, dyed by XLY, 4) absorption of PCA, dyed by XLY, 5) absorption of initial PCA.

Similar results were obtained during investigation of spectra of excitation, absorption and luminescence of films and solutions of dyed and undyed PCA samples, and also on using other dyes- stabilizers of this series.

Absence of dyed PCA in fluorescence spectrum, containing 1% ($1,8\cdot10^{-2}$ $^{mole}/_{kg}$) of the dye from polymer mass of the peak $\lambda_{max}{}^{em}$=425nm (fluorescence peak of initial PCA), may be explained by large fluorescence intensity of the dye itself. Since it is impossible to judge suppression of PCA fluorescence at such relatively high dye concentration, measurements of fluorescence at lower concentrations have been carried out. PCA films with 10-20 μm of thickness, containing $7\cdot10^{-5}$-$1,5\cdot10^{-2}$ $^{mole}/_{kg}$ of the dye were obtained for these measurements. There were read spectra of fluorescence of films (initial and thermoprocessed) and polymer solutions in the mixture of formic (85%) and tri fluoro-acetic acids in relation 1:1.

During check experiments it has been shown that the solution being used does not influence spectral characteristics of tested samples in investigated region of wave-lengths.

Spectra of PCA films fluorescence with different content of XLY compounds are given in Figure 1. 14. As it is seen from the figure intensity of the peak $\lambda_{max}{}^{em}$=425nm decreases with the rise of dye concentration, but at dye concentration $0,7\cdot10^{-3}$ $^{mole}/_{kg}$ spectrum character changes – there takes place the shift of the peak of PCA fluorescence into the region with longer waves (470-490nm). Averaging of luminescence peak of initial PCA and its dyed analog takes place. Shift of the peak into the region with longer waves may be explained not only by additive absorption, but by the beginning of realization of energy transfer from PCA to the dye and its simultaneous deactivation by luminescence of the dye itself.

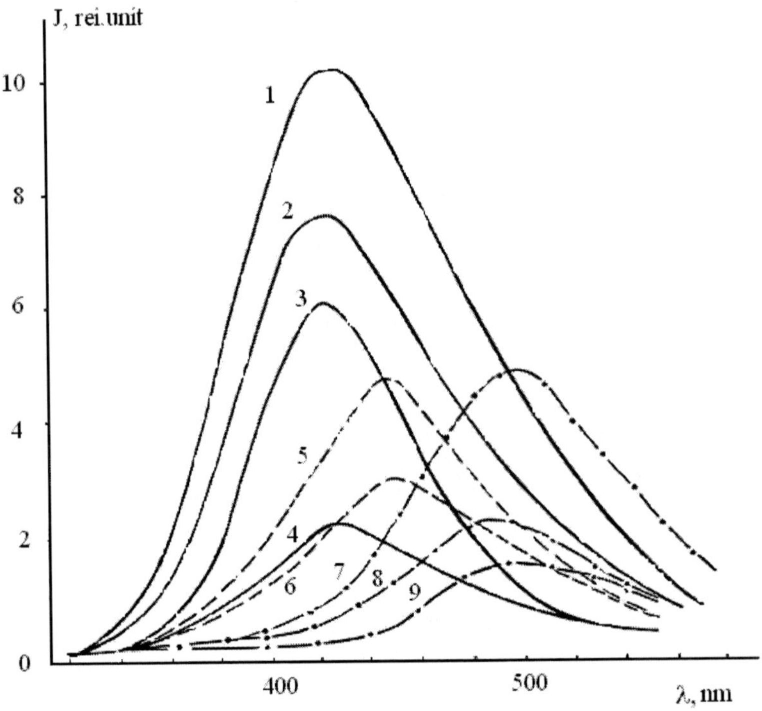

Figure1.14. Luminescence spectra of PCA films: initial (I) and dyed by XLY compound (2-9). Conservation of XLY (mole/kg); 0,7·10-4 (2); 1,5·10-4 (3); 3,5·10-4 (4); 0,7·10-3 (5); 1,5·10-3 (6); 3,5·10-3 (7); 0,7·10-2 (8); 1,5·10-2 (9).

Complete disappearance of PCA luminescence peak and appearance of true dye glow is being observed at dye concentration of $3{,}5\times10^{-3}$ mole/kg and higher, moreover its further content leads to decrease of dye luminescence intensity. This may be classified as the process of concentration of self-supression, which is a positive factor, retarding consumption of the dye in the process of irradiation.

Additives, having really used concentrations, provide protective effect, though luminescence on the wave-length 420-430 nm has been already suppressed by them, so there are no reasons to connect protective effect only with the process of suppression. This means that polymer protection over shielding effect needs explanation through chemical action of additives being used. Study of this photooxidation process must be the subject of separate invesrigation.

The totality of these results shows, that:

(1) obtained protective effects does not come to shielding effect: used dyes by no means play the role of suppressors of polymer excited states, however, protection over shielding cannot be explained mainly by processes of suppression;
(2) since literature data, related to used irradiation conditions, do not allow to consider protective role of possible effect of additives on supermolecular polymer structure, then one may come to a conclusion that there is considerable contribution of chemical inhibition of polymer photooxidation by applied additives in observed protective effects;
(3) chemical mechanism of protective action of applied additives needs additional investigation, necessary for scientifically well-founded selection of much more effective light-stabilizing compositions.

1.7. Effect of Long-Wave Irradiation on Dyed PCA-Fibre

Since PCA-products are subjected to influence of wide spectrum of light radiation then it is interesting to study the effect of long-wave irradiation on PCA-fibres dyed by bis-aroilenbenzimidazole compounds. It should be noted that in the region with λ_{MAX}^{abs} more than 480 nm additives XLIY-XLYI practically do not absorb acting light.

Data on dependence of viscosity change of undyed PCA-fibres and dyed by additive XLY on the wave-length of acting light are given in Figure1.15. As it is seen from it, additive XLY shows both stabilizing and sensibilizing properties depending on wave-length of acting light.

During irradiation of PCA-fibre dyed by additive XLY by full spectrum of day-light lamp LDTs – 80 it has been found that with increase of exposure time intensity of additive absorption at λ_{abs}=420 nm decreases and already after 8 days of irradiation the dye is consumed practically completely.

Taking as a basis these facts one may come to a conclusion that total effect of suppression, shielding and sensibilizing is realized in ultra-violet region. They act as sensibilizers outside the limits of absorption spectrum of bis-aroilenbenzimidazole itself, where there are practically no effects of shielding and suppression. The last conclusion was proved at relative study of the effect of additives XLY and XLYII, similar according to their structural features, but absorption of the additive XLYII takes place in the region with longer wave with λ_{abs}=460-560 nm.

As it is seen from Table 10 stabilizing activity of PCA-fibre dyed by the additive XLYII is observed during long-wave irradiation, and at the same time this index is twice lower in a sample dyed by the additive XLY.

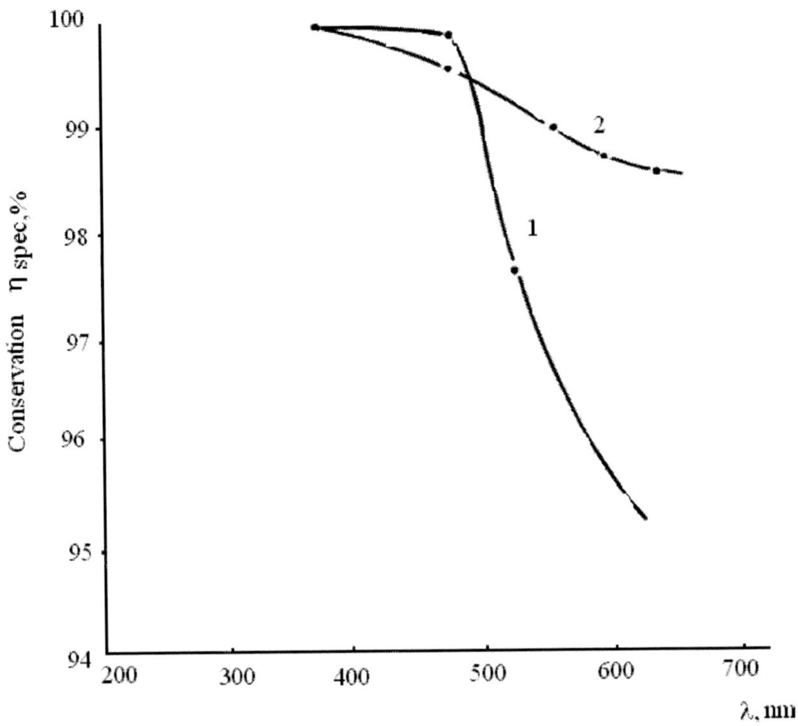

Figure 1.15. Conservation of specific viscosity of dyed (I) and undyed (2) PCA-fibre at different wavelength of irradiation and exposure time of 45 days.

Table 10. Viscosity characteristics of PCA-fibres

Number	Source of irradiation	Polymer sample	Specific viscosity of irradiation		% of conservation ή specific
			Before irradiation	After irradiation	
1.	LDTs-80	PCA-initial	2,2	1,2	53,2
2.	-//-	PCA + XLYII	2,2	1,7	77,5
3	-//-	PCA + XLY	2,2	0,7	39,4
1.	PRK-2	PCA-initial	2,2	0,9	43,2
2.	-//-	PCA + XLYII	2,2	1,2	56,0
3	-//-	PCA + XLY	2,2	2,0	94,3

In ultra-violet region the picture is a little different – stabilizing effect of the dye XLYII is extremely small in comparison with initial PCA-fibre and dyed by the dye XLY.

Hence, aroilenbenzimidazoles do not show protective action and even become sensibilizers outside the limits of absorption regions of acting light and all this requires individual approach to the synthesis of additives for PCA depending on on the range of acting light during the use of products from this polymer.

1.8. THERMOOXIDATIVE DESTRUCTION OF MODIFIED PCA

PCA warming-up in the presence of oxygen or in the air leads to great changes in polymer chemical composition, which is accompanied by the loss of positive service properties of PCA materials. These changes allow to registrate complex thermogravimetric method of analysis.

So, investigations of effect of bis-(1', 8' – benzoilen – 1, 2 - benzimidazoles) and bis – (1,8 – naphthoilene – 1,2 – benzimidazoles) on thermooxidative PCA destruction in conditions of dynamic and isothermic heating have been carried out in this connection.

Data of complex thermogravimetric analysis showing that introduction of bis-aroilenbenzimidazole derivatives into PCA shifts temperature of beginning of PCA depolymerization by 10-25 ^{0}C into the region of higher temperatures are given in Table 11.

Table 11. Thermal characteristics of PCA fibre dyed by derivatives of bis-aroilenbenzimidazole

Fibre sample	T melting, ^{0}C	T-beginning of decomposition, ^{0}C	Temperature of mass loss,		
			5%	10%	25%
PCA – initial	216	325	347	368	407
PCA + additive XL	216	338	356	380	410
PCA + additive XLI	217	352	358	385	412
PCA + additive XLII	215	333	355	379	408
PCA + additive XLIII	215	335	350	373	405
PCA + additive XLIV	218	341	363	386	415
PCA + additive XLV	217	364	365	390	418
PCA + additive XLVI	217	336	356	382	412

Similar effect is probably connected with structural changes in PCA, taking place at addition of compounds XLY-XLYI into polymer and their possible inhibiting effect on PCA thermooxidative destruction.

Performing differential thermal analysis of the processes taking place in the field of polymer melting (T=215-220 ^{0}C) has disclosed great differences in dyed and undyed samples. Initial polymer (Figure 1.16) has wide peak of exoeffect in temperature range of 160-215 ^{0}C before melting peak (endoeffect at 220 ^{0}C), which is practically absent in its dyed analogs.

Such difference is explained by the effect of XLY-XLYI compounds on the ordering of PCA supermolecular structure during the period of its formation as it comes from the results of X-ray diffraction analysis. PCA fibres, containing XLY-XLYI compounds, in the range of considered temperatures have marked melting peak (endoeffect at 218-220 ^{0}C) [134]. Clear character of melting peak shows narrow distribution of crystal part of the polymer according to the sizes of crystallites [135].

Investidations in isothermal conditions have been carried out in the air at the temperature 150-200 ^{0}C. Data of kinetics of samples mass loss during warming up show, that initial sample (Figure 1.17, curve I) detects the greatest mass loss: undyed sample loses about 20% for 6 hours and the one dyed by XLY compound – 31%. Induction period takes place in dyed samples, but it is absent in initial PCA. Similar effect was observed in the work [136] during investigation of dyes effect on thermal ageing of modular polymer.

Hence, introduction of bis-aroilenbenzimidazole derivatives into PCA increases polymer thermal stability at warming up in the air, moreover the greatest effect is displayed in compounds containing Ar-O-Ar and Ar-CH_2-Ar in a bridge group. On the basis of obtained results on photo- and thermooxidative PCA destruction one may come to a conclusion that compounds possess light- and thermoprotective action, being displayed in increasing photo- and thermal stability of the products from PCA.

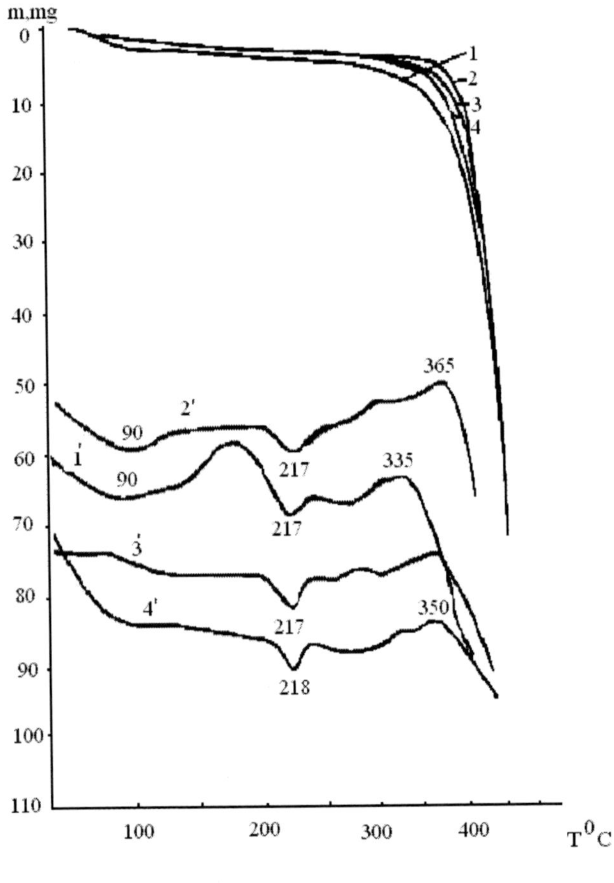

Figure 1.16. Thermograms of unstabilized (I, I^1) and stabilized compound XLY (2, 2^1), XLYI (3, 3^1) and (4, 4^1) of PCA-fibre in the air.

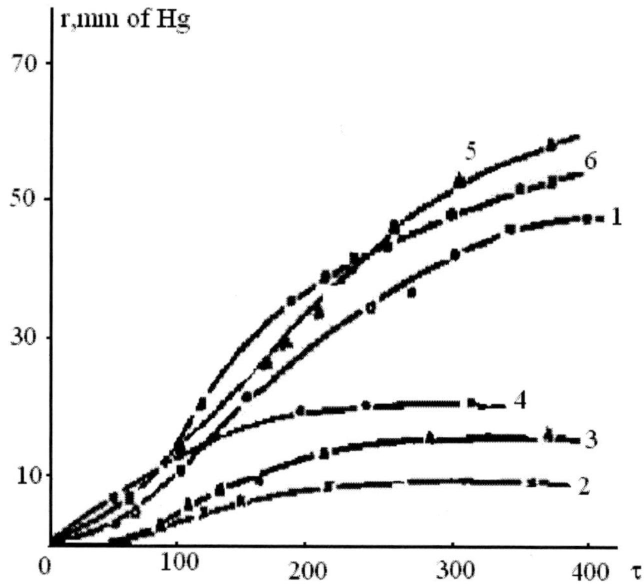

Figure 1.17. Kinetic curves of mass loss of unstabilized (I) and stabilized compounds XLIII-(2), XLIY-(3), XLY-(4), XYI-(5) of PCA-fibre in isothermal conditions at T=200 ^{0}C.

It was shown earlier that (diphenylamine-p-tolylive-cyanate) urea (DTCU) may be used as antioxidant of thermo- and photooxidative destruction of AC. It was interesting to use DTCU for increasing antioxidative stability of PCA. With this purpose small amount of DTCU were introduced into **ε** – caprolactam directly before its polycondensation, which was carried out according to well known methods [132]. Content of DTCU additives in polycondensation mixture was 0,5 – 3% from caprolactam mass.

Data on effect of DTCU concentration on viscosity characteristics and change of PCA molecular mass before and after ultra – violet irradiation are given in Table 12.

Table 12. Dependence of viscosity characteristics of PCA on DTCU content before and after ultra-violet irradiation

DTCU content in % from PCA mass	Specific viscosity		% of conservation, η specific
	Before irradiation	After irradiation	
0	0,68	0,54	80
0,5	0,66	0,62	94
1,0	0,62	0,60	97
2,0	0,76	0,70	82
3,0	0,56	0,52	93

As it seen from the Table DTCU additives (0,5-2% from the mass of ε-caprolactam) do not exercise influence on polymer viscosity characteristics and, hence, on its molecular mass.

If one may consider that during PCA photooxidation the rate of formation of joints between polymer molecules is much lower than during photolysis in vacuum [4], then little change of specific viscosity of irradiated PCA, containing DTCU 0,5-2% from polymer mass, may be related to stabilizing action of the latter. Similar effect is probably connected with shielding action, as in the case of AC.

Kinetic curves of oxygen absorption at constant temperature 200 ^{0}C were obtained to discover the effect of DTCU additives on thermooxidative stability. Curves of oxygen absorption by modified and initial PCA before and after ultra-violet irradiation are given in Figure 1.18. Analysis of curves (Figure 1.18) shows unified mechanism of PCA oxidation. A bit larger oxygen absorption by initial PCA is probably connected with larger specific internal surface of polymer, whereas surface of modified polymer is filled by DTCU additives. Hence it follows that ultra-violet irradiation increases resistance of modified PCA to thermooxidative destruction. Similar effect was observed earlier [137] during studying thermooxidative destruction of natural silk fibration. This effect may be explained both as antioxidative activity of DTCU additive [138], and activating effect of ultra-violet irradiation on DTCU, which is quite proved by data on kinetics of oxygen absorption by DTCU itself before and after irradiation (Figure 1.19). Activating effect of ultra-violet light is probably due to the fact that at severe irradiation paramagnetic centres are formed in DTCU which is proved by electronic paramagnetic resonance (EPR) spectra (singlet on EPR curve with g=0,002, showing presence of radicals). EPR signal was not observed in unirradiated DTCU samples.

Thus, one may come to a conlusion that PCA modification by introducing DTCU into polymer and irradiation by ultra-violet light leads to increase of its thermooxidative stability. Hexaazocyclanes (HAC) and their salts (iodine and bromine) ($LV + 2J^-$), ($LV + 2Br^-$) may be other possible thermostabilizers of PA. Taking into account the possibility of long utilization of polymers at high temperatures we studied kinetics of PA destruction during isothermal heating at 200 ^{0}C during 6 hours in the air.

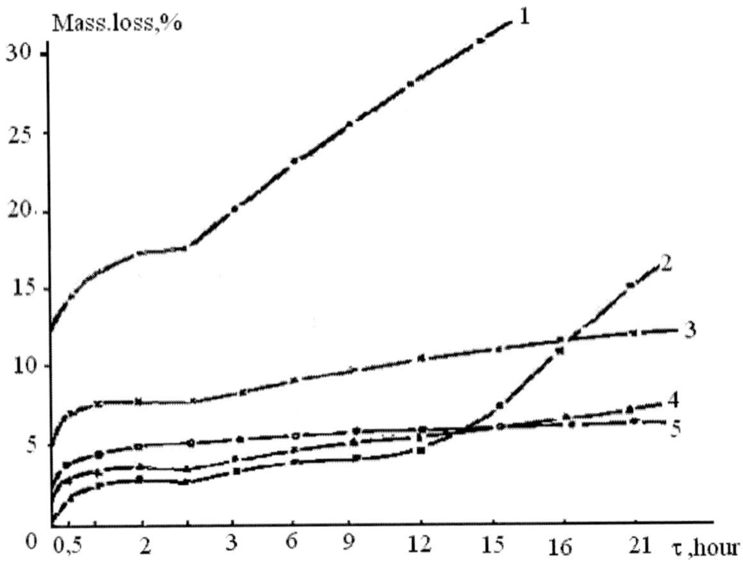

Figure 1.18. Oxygen absorption by PCA with DTCU additives before (1, 5, 6) and after (2-4) irradiation. DTCU content from PCA mass: 2-0,5%, 3-1%, 4-3%, I-0,5%, 5-1% (before irradiation).

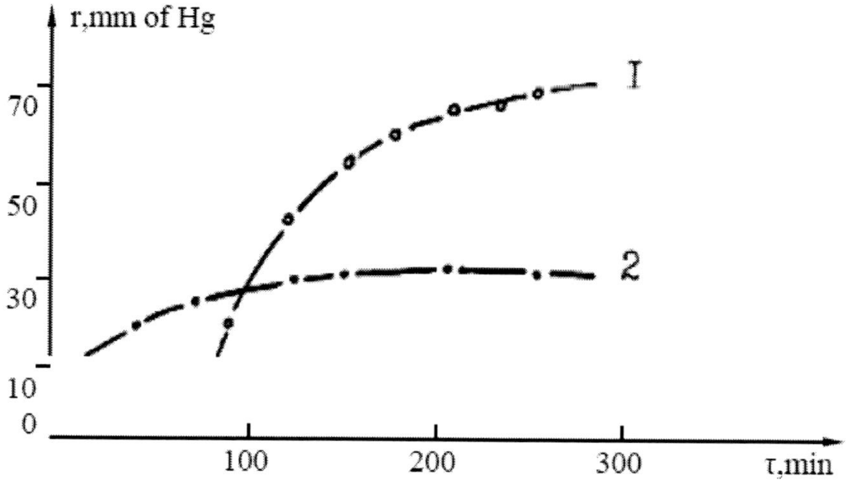

Figure 1.19. Oxygen absorption by DTCU before (I) and after (2) irradiation.

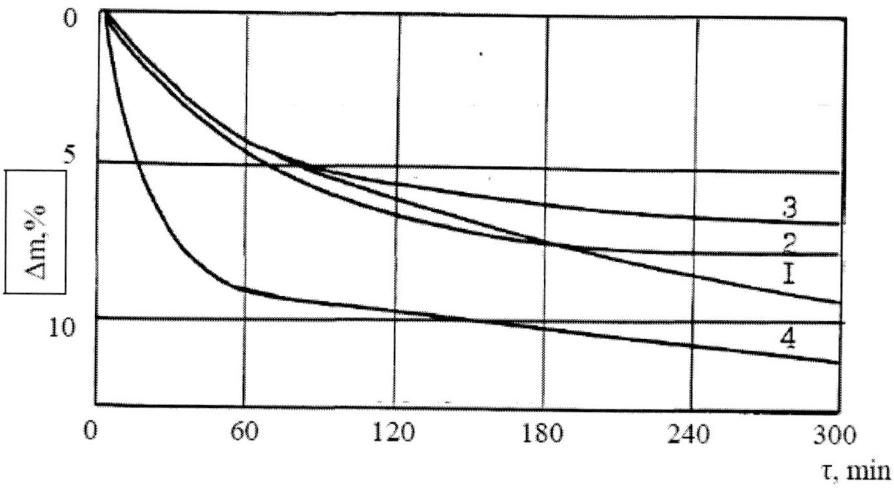

Figure 1.20. Dependence of PCA mass loss on duration of heating at 200 ^{0}C of stabilized by: I-LV; 2-(LV+J ^{2}); 3-(LV+Br ^{2}); 4-initial.

Stabilizing activity of HC is observed at thermooxidative destruction (Figure1.20). Moreover, in this case the following sequence of thermostabilizing activity is characteristic:

$(LV+2Br^{-})>(LV+2J^{-})>(LV)$

These conclusions were proved by the data of complex thermographic analysis. Temperature of intensive destruction of modified polymers is higher, than in initial, by 15-30 ^{0}C. Going on with earlier works [139] on investigation light and thermal stability of polymers with additives of carbazolesulfonamides, thermooxidative stability of PCA dyed by azo dyes of the following structures was studied:

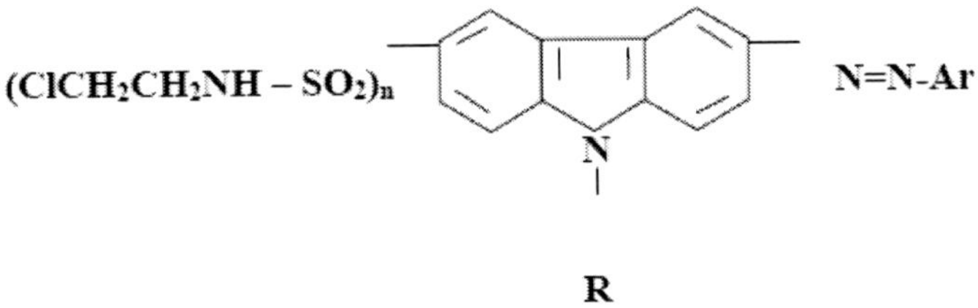

Where R=H, CH_3, n=1, 2, 3

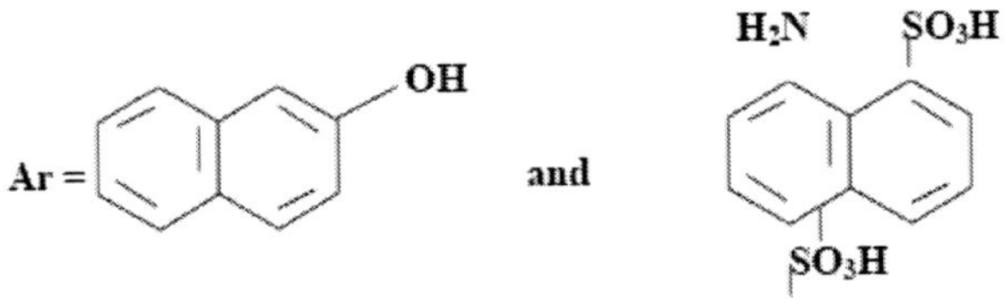

With this purpose there were taken derivatograms of dyed samples in comparison with initial undyed ones. It was noted, that similar dyeing does not greatly effect on the process of softening of fibre-forming polymer. However, in the range of temperatures 210-220 ^{0}C, that is in the field of PCA melting there are some disagreements between initial and dyed somples (Figure 1.21).

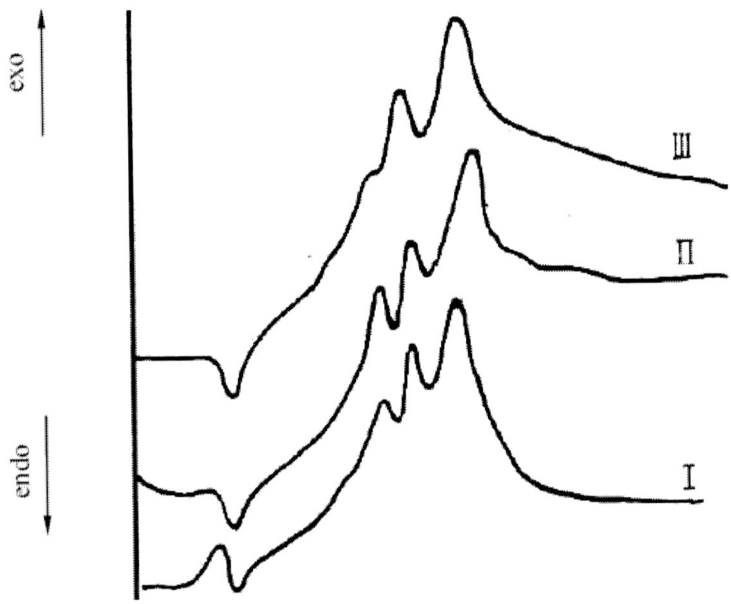

Figure 1.21 PCA DTA curves: I-initial, II-dyed by LXI, III-dyed by LXII.

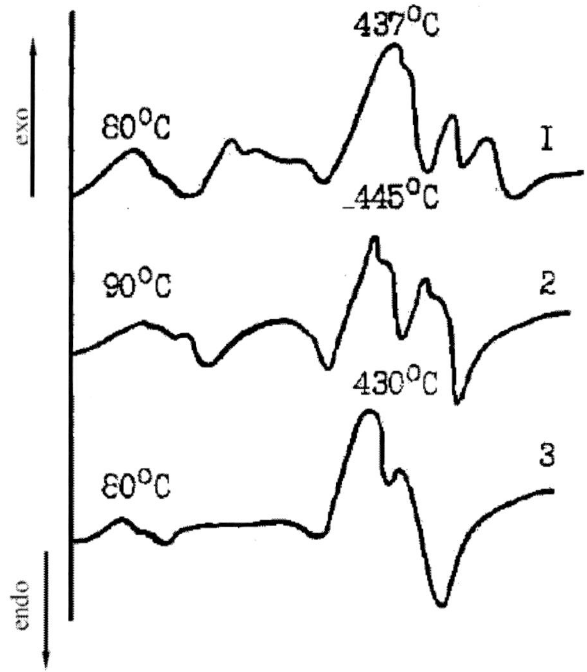

Figure 1.22 DTA curves in the air: I-undyed kapron; 2-kapron dyed by active golden yellow RCr covalently linked; 3-kapron dyed by hydrolyzed golden yellow.

Exoeffect at 210 ^{0}C appears in initial sample before melting (endoeffect at 220 ^{0}C), which is probably connected with polymer crystallization being absent in dyed fibre.

May be this difference is connected with the fact that the dye is situated in amorphous regions of PCA and is linked with functional groups by covalent bonds, which, as a result, limits the possibility of polymer crystallization.

So, one may come to a conclusion that proposed dyes possess high light protective and thermooxidative action.

Complex thermographic investigation of PCA-fibre samples, containing the dye linked with polymer by both covalent forces (active dye-LXI) and chemosorption forces (hydrolyzed form of the dye) – LXII has been carried out.

Kapron was dyed by dichlortriazine dye – LXI and its deactivated dioxytriazine form. Dye content on the fibre did not exceed 1% from polymer mass.

The first small endoeffect, accompanied by a little mass loss within temperature range 80-90 ^{0}C, is connected with the loss of absorbed moisture. Endoeffect at 203-218 ^{0}C without loss of weight corresponds to melting of PCA crystal regions. Further thermal effects are connected with oxidative destruction of polymer. As it is seen from Figure 1.21, there are no sharp differences in thermal effects both in initial and dyed PCA.

Quite another picture is observed during investigation of DTA curves, obtained in inert atmosphere (Figure 1.22). Thermal effects of moisture loss and melting are observed correspondingly at 80 and 210 ^{0}C. If deep effect is observed in the air in temperature range 430-445 ^{0}C, accompanied by loss of mass, then a number of endoeffects are observed in argon in temperature range 420-480 ^{0}C, also accompanied by loss of weight.

While considering DTA curves (see Figure 1.22 and 1.23) it should be meant that thermal decay of PCA (endotherm) occurs in inert atmosphere, whereas reaction of PCA oxidation (exotherm) takes place together with it in the presence of air oxygen according to the mechanism, proposed in the work [140].

Curves of mass conservation at temperature rise (Figure 1.24) show that initial PCA (curve 4) is more stable in inert atmosphere, than in the air (curve 2). At the same time quite another picture is observed for dyed samples. Here, the greatest effect of PCA protection is observed at warming up in the air. So, initial PCA loses 60% of original mass at the temperature of 400 ^{0}C, while the dyed one – 18%. And it should be noted that PCA, dyed by covalently linked dye, possesses higher resistance to thermooxidative destruction, than PCA, dyed by the same dye without forming covalent bond.

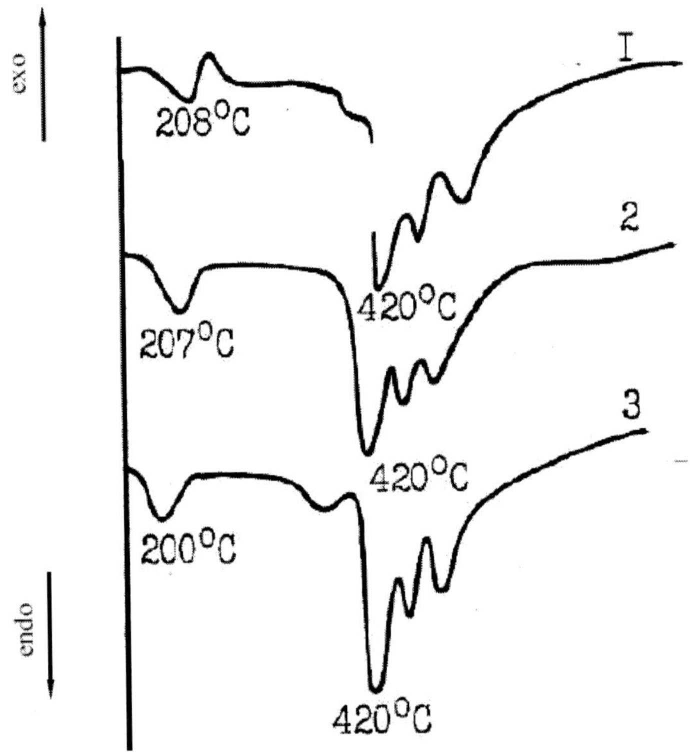

Figure 1.23. DTA curves in argon, I- undyed kapron, 2- kapron dyed by active golden, yellow RCr covalently linked 3- kapron, dyed by hydrolized golden-yellow.

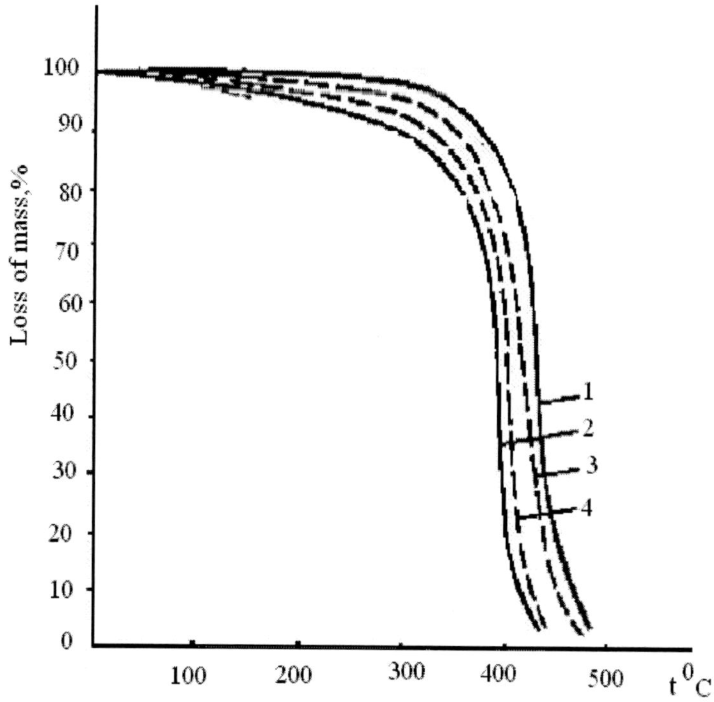

Figure 1.24. Curves of dependence of mass conservation on temperature 1-kapron, dyed by golden-yellow RCr, covalently linked (in the air), 2- undyed kapron (in the air), 3- kapron, dyed by active golden-yellow RCr (in argon), 4- undyed kapron (in argon).

Chapter 2

STABILIZATION AND MODIFICATION OF CELLULOSE DIACETATE

2.1. Modern State of Investigations of Photochemical Destruction of CDA

Cellulose and its derivatives – cellulose acetate – are renewed polymers, that, together with the whole complex of valuable and indispensable properties, defines continuous growth of their production.

Acetate fibres differ from cellulose fibres in light and thermooxidative stability, as the presence of ester groups decreases stability of molecular structure, owing to which destructive processes begin at much lower temperatures and weak energy effects.

Since macromolecules of cellulose acetate are constructed on the basis of cellulose then mechanism of photodestruction of these polymers may be considered as general.

Many summarizing works [141-147], published from 1962 to 2000, and are devoted to the questions of photochemistry. In this survey there are works, which are not included into above-mentioned literature surveys, and publications of the last years.

The most important energetic factor, which photodestruction of cellulose and its derivatives depend on, is intensity of irradiation and wave length. Destruction of cellulose and its derivatives under atmospheric conditions, proceeding as a result of photochemical reaction, on the whole takes place under the action of ultraviolet rays with λ=200-360 nm. Since cellulose contains three types of chromophore groups – hydroxyl, acetate and semiacetate and also aldehyde – then it is considered that light absorption in the region of 250-300 nm is caused just by them. At the same time some authors, bringing the possibility of light absorption by acetal chromophore in question [148], have put forward the supposition [149] that photochemically active centres in cellulose materials, containing carboxyl and hydroxyl groups, may be molecular complexes between these groups, connected by the system of hydrogen bounds with definite energy of interaction. Disproportion of intermolecular bonds, providing fixation of excited state in cellulose matrix takes place in such complexes at their excitation. Thus, there are many different hypotheses, often contradicting each other, about the effect of chromophore groups on light absorption by cellulose.

There are many data about the nature of free-radical particles, being formed at irradiation of cellulose by ultraviolet light. Since, being formed products of phototransformation are highly mobile and easily undergo further transformations, method of electron-paramagnetic resonance (ESP) is one of the most effective for these particles identification. Critical analysis of a great number of works on EPR spectra interpretation is quite fully given in surveys [150-152]. More than 20 different radicals are being formed at ultraviolet irradiation as a result of break of practically all bonds C-C; C-H; C-O. Main types of macroradicals, with indication of atom and groups of atoms after removal of which these macroradicals are formed, are presented in scheme 1. Formation of five lowmolecular radicals: OH, CHO, H, CH_2OH, CH_3, is also marked here.

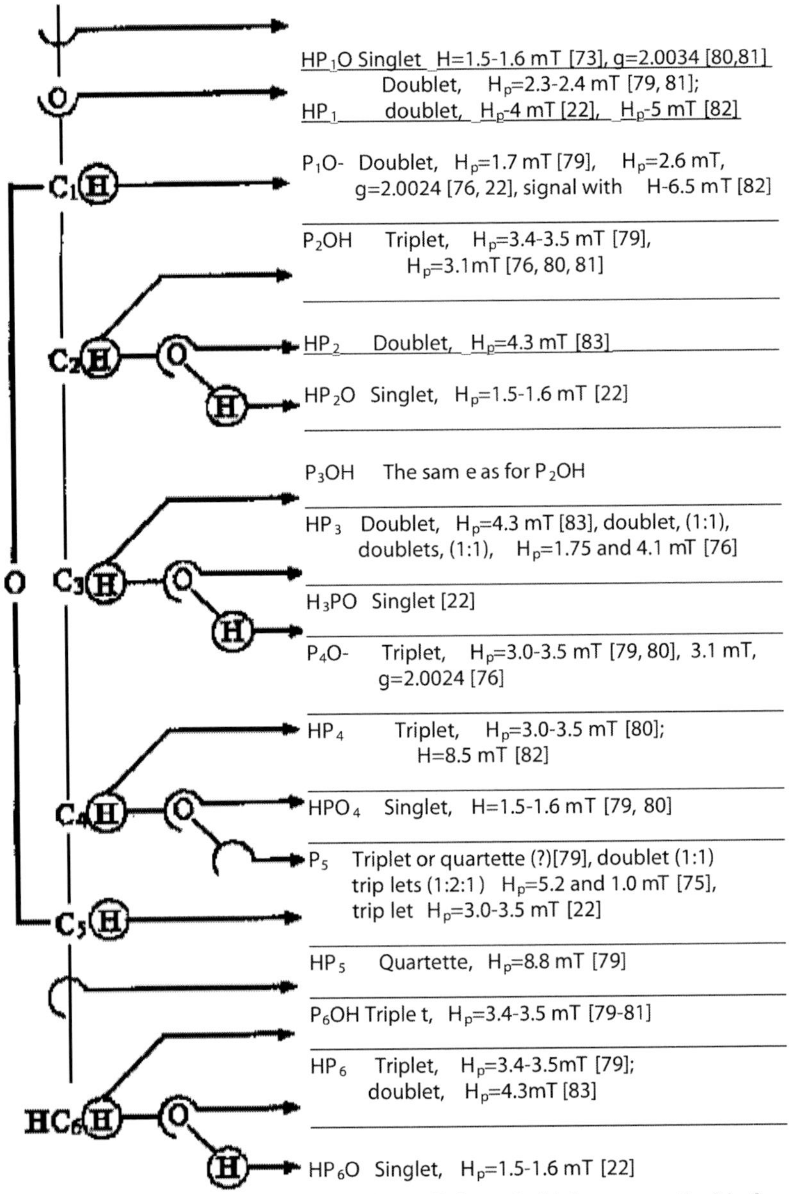

* The arrow indicates atom or groups of atoms, after splitting of which macroradical is formed:

* The index shows the number of atom C, on which valency is localized, or atom C, being the nearest to the place of free valency localization.

Scheme 1.

Composition and properties of radicals, being formed under the light action, depend on conditions of experiment (temperature, light intensity, spectral composition of light and soon). Besides, EPR spectra of some radicals depend on cellulose structure. That is why EPR spectra of cellulose have complex character. Problem of these spectra interpretation has not been completed yet, and identification of a number of radicals is debatable. Analysis and conclusions, made while discussing investigations on cellulose photodestruction, greatly facilitate the approach to similar processes understanding, taking place at light-ageing of di- and triacetate of cellulose, though they have their own features [153, 154].

The process of cellulose acetate oxidation under the action of light energy proceeds according to chain mechanism with formation of free radicals and different gaseous products [155, 156, 157]. Depending on conditions of irradiation proportion of rates of separate stages of chain process changes, but unfortunately, kinetic parameters of this process are not defined and this does not allow to judge the length of the chain of cellulose acetate (CA) photooxidation.

Investigation of the mechanism of photo- and photooxidative destruction has shown [141, 154 158] that intrinsic viscosity decreases at photodestruction of cellulose acetate, content of combined acetic acid also decreases and accumulation of carbonyl groups takes place. There have been identified six main volatile products: CH_2=C=O; CO; CO_2; H_2; H_2O; CH_3COOH, moreover acetic acid is the main product [159]. In some authors opinion, break of acetal bond 1-4 and opening of pyranose cycle according to C_1-C_2 happen at photodestruction of cellulose acetate.

It has been stated that the first stage of chain process, developing at light action on cellulose acetate, is appearing of free radicals [2]. Phototransformation of radicals, being formed, has been discussed in details in works [2, 148. Probably, breaks of bonds may take place according to the following mechanism:

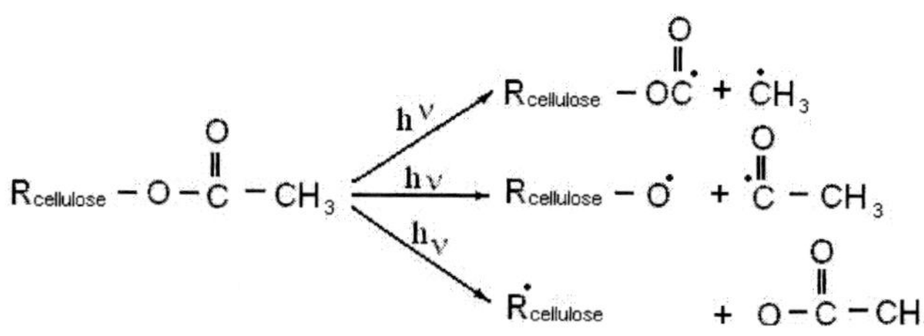

Besides, the possibility that acetoxyl radical CH_3COO* may be formed at trans-splitting of acetal groups from cellulose acetate is not ruled out:

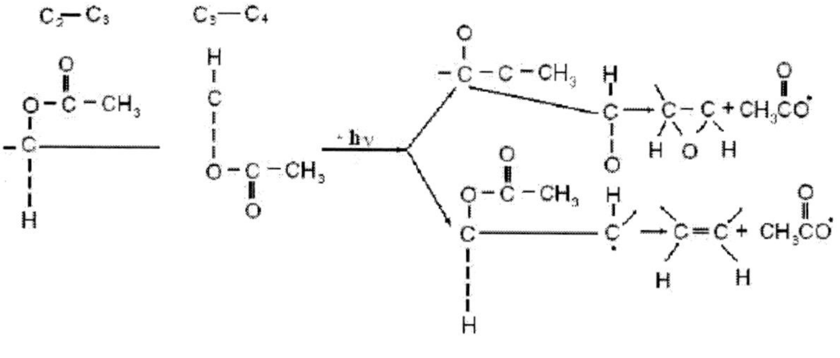

It should be noted that formation of ketene and acetic acid is not observed at photolysis of glucose and cellulose.

So, one may come to a conclusion: there is no unity of views of researchers regarding photodestruction of cellulose and its derivatives, and for better understanding the mechanism of phototransformation it is necessary to take into account that while studying kinetics of phototransformation one should consider the following factors: effect of supermolecular structure and initialing or inhibiting action of impurities.

2.2. About the Mechanism of Photooxidative Destruction of Cellulose Acetate

As the main product of CA photodestruction is acetic acid, formed as a result of the break of bonds C-O-C at carbon atoms in positions 2, 3 or 6 [2], it is suggested that break of these bonds run in direct photolysis with splitting radical AcO* out, which in further reactions breaks atom H from polymer and is transformed into acetic acid [2].

However, being formed acetyl radical (AcO*) is not stable and easily decomposes with formation of CO_2 [160]. That is why, kinetics of accumulation of acetic acid and radicals at initial stages of cellulose diacetate (CDA) photolysis has been studied for full understanding of the mechanism of glucoside bonds breakage and formation of acetic acid. Typical curves of acetic acid accumulation at CA films irradiation are given in Figure 2.1.

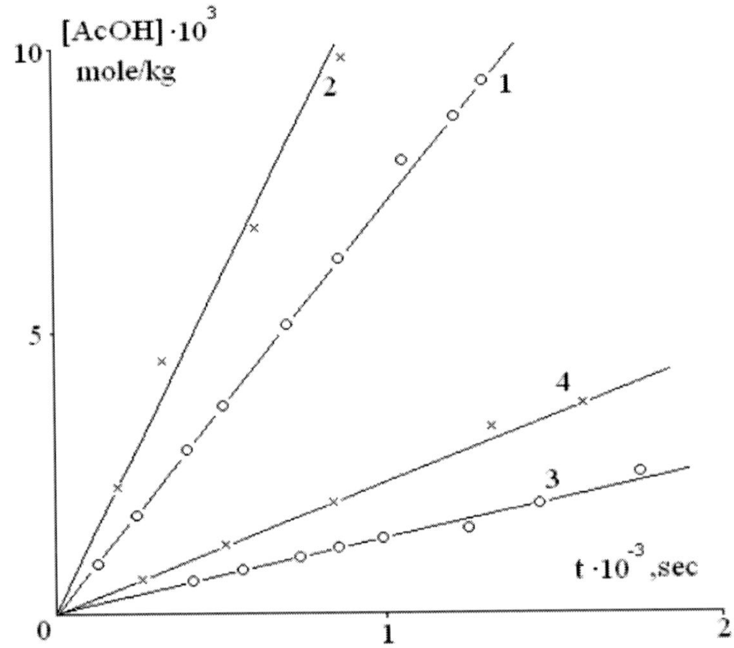

Figure 2.1. Kinetic curves of acetic acid accumulation at CA films irradiation at 25°C in vacuum (1,3) and in the air (2,4); light intensity is 10·1014 (1,2) and 3,6·1014 quant/cm2·sec (3,4).

As it is seen from Figure 2.1 constant rate of acetic acid accumulation is stated soon after the beginning of irradiation. The rate of the process in the presence of oxygen of the air is only 1,5-2 times larger than is vacuum. This agrees with the data of [161] and shows, that reaction of acetic acid formation in the absence of oxygen plays an important role at CA photodestruction.

Evaluation of quantum yield of acetic acid formation at light intensity (λ=254 nm) $I=1\cdot10^{15}$ quant/cm^2·sec gives the value Φ=0,02 close to the value of quantum yield of ester groups destruction (Φ=0,015), measured in similar conditions [162]. Close value of quantum yield of acetic acid formation Φ=0,01 has been obtained in the case of polyvinyl acetate photolysis at lower intensity – $I=5,7\cdot10^{14}$ quant/cm^2·sec [163]. Data on effect of temperature on the rate of acetic formation are given in Table 1. Activation energy of the reaction of acetic acid formation at CA irradiation, calculated according to these results, is E_φ=9,66 kJ/mole, which is characteristic for photoprocesses.

Table 13. Effect of temperature on the rate of acetic acid formation at CA ageing in vacuum in darkness (W_T) and under light action (W_φ)

T, K	$W_T\cdot10^6$ mole/kg·sec	$W_\varphi\cdot10^6$ mole/kg·sec
253	0,01	0,46
298	0,001 – 0,01	1,15
366	3,90	2,12

W_φ has been got by subtraction W_T from the general rate of the process under light action.

Let's note, that acetic acid formation runs not only under the light action, but in darkness too. Evaluation of activation energy of dark process gives the value E_T= 42-55 kJ/mole. Measurement of stationary rates of acetic acid formation in the wide range of light intensities shows that dependence of the process rate on light intensity has complex character (Figure 2.1.). At large intensities ($I>1\cdot10^{14}$ quant/cm^2·sec), W - $I^{1,5}$; at low intensities ($I<1\cdot10^{13}$ quant/cm^2·sec), W – $I^{0,5}$ and only at intermediate values $I=(0,1\text{-}1)\cdot10^{14}$ quant/cm^2·sec – the rate of the process is proportional to light intensity.

Presence of induction periods on kinetic curves of acetic acid accumulation (Figure 2.1.) and existence of higher, than a unit, order of the reaction according to light intensity (Figure 2.2.) does not agree with the hypothesis on acetic acid formation in the primary process of photolysis of ester bond.

Additional confirmation of this conclusion has been got during the experiments on measurement of dependence of the rate of acetic acid formation on light intensity on one and the same film. It has been found that if the film is being irradiated changing from lower intensity to greater one, then W – $I^{1,5}$ (Figure 2.3, strait line 1). And if during film irradiation intensity is successively being decreased, then W – I (Figure 2.3, strait line 2).

These data show that acetic acid is being formed at photolysis of intermediate product, concentration of which depends on light intensity. Such intermediate product may be radicals, photolysis of which leads to acetic acid formation. Scheme 2 illustrates this on the whole:

CA + hv → R*
P + hv → AcOH + R*
R* + R* → products

Scheme 2.

Found hysteresis of the dependence of photolysis rate on light intensity is directly connected with radicals concentration. At successive decreasing of light intensity radicals concentration, during the experiment, has remained invariable and equal to maximum stationary concentration of radicals, achieved at the greatest light intensity.

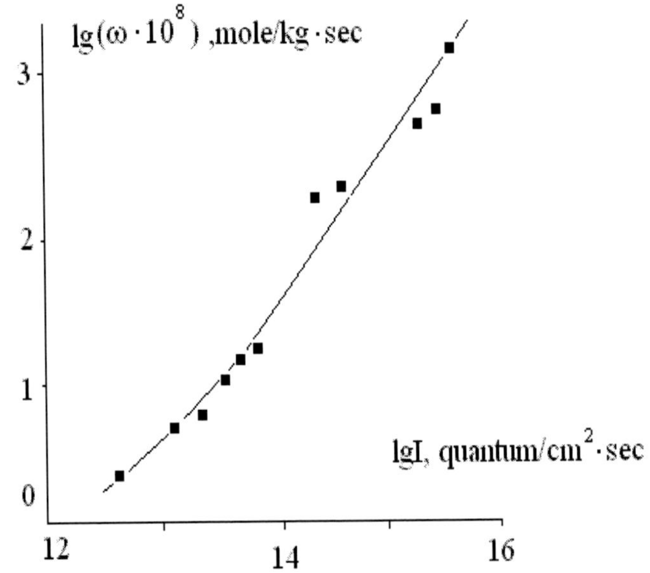

Figure 2.2. Dependence of the rate of the acetic acid accumulation on light intensity at 25°C in vacuum.

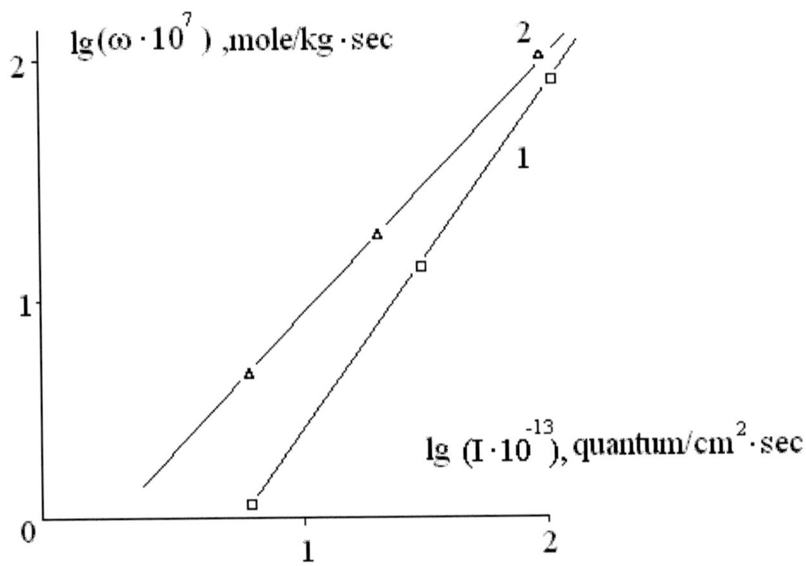

Figure 2.3. Effect of light intensity on the rate of acetic acid accumulation at successive increase (I) and decrease (2) of light intensity at irradiation of one and the same CA film in vacuum at 25°C.

Accordingly, the rate of acetic acid formation in these, specially chosen, conditions was proportional to light intensity at constant concentration of radicals. And if there is possibility for radicals to be lost and then irradiation is renewed, changing from lower intensity to higher one, then stationary concentration of radicals will increase proportionally to $I^{0,5}$ and the rate of photolysis – proportionally to $I^{1,5}$. Hence, the rate of photolysis is proportional to light intensity and concentration of radicals. These data show that acetic acid is formed at photolysis of radicals and in connection with this radical processes at CA photolysis have been studied.

2.3. Kinetics of Radicals Accumulation

Quantitative study of kinetics of radicals accumulation has required solution of auxiliary problem – definition of the rate of photoinitiation W_{in}. In the case of solid-phase reactions there are experimental difficulties in solution of this problem. Measurement of W_{in} according to consumption of inhibitor is complicated by possible photochemical reactions of inhibitor itself and specific solid-phase effects of "kinetic stop" type and so on [10]. Measurement of W_{in} according to initial rate of radicals accumulation is also tactless in solid polymer, as the latter may be much lower than W_{in} [164].

1,4 – trichlormethylphenylene (TCMP) was chosen as photoinitiator for solution of the set problem. Under light action TCMP splits atom of chlorine, which possesses high reactivity and easily breaks hydrogen atom off the polymer [165]. So, macroradical of polymer and molecule of chlorine hydrogen are formed in primary act of TCMP photoinitiation. That is why formation of chlorine hydrogen, which is easily measured by pressure gauge, is peculiar counter of polymer macroradicals being formed.

It has been found that the rate of chlorine hydrogen formation at TCMP photolysis in CA in the wide range of light intensities and TCMP concentrations (0,01-0,2 mole/kg) is proportional to light intensity and TCMP concentration. Quantum yield of hydrochloric acid, defined from these data, was equal to Φ_{HCL}=0,028. Data on kinetics of accumulation of HCl and macroradicals at CA photolysis with TCMP addition are given in Figure 2.4. The rate of HCl formation (and also W_{in}) remains constant during the whole period of films irradiation that is why, determination of stationary concentration of radicals is caused by determination of balance between reactions of formation and destruction of macroradicals.

According to the data of Figure 1.4 a, b the rate of photoinitiation decreases by 20 times at light intensity decreasing by 20 times. This shows that in experiment conditions W_{in} – I, as it was to be expected. Stationary concentration of radicals $(R^*)_{st}$ decreases from 0,001 (Figure 2.4a) to 0,0002 mole/kg (Figure 2.4 b), that agrees with the law $(R^*)_{st}$ – $I^{0,5}$ which is usually characteristic in the case of quadratic destruction of radicals. The rate of photoinitiation is equal to the rate of all radicals formation at polymer photolysis. Initial rate of radicals accumulation is equal to such one only for those radicals, which in the conditions of experiments are registrated by EPR method. It may be lower than the rate of photoinitiation, if it is impossible to registrate the part of radicals because of high constant of the rate of their destruction [166].

According to the data, given in Figure 2.1., decrease of light intensity by 20 times leads to decrease of the rate of observed radicals formation by 80 times.

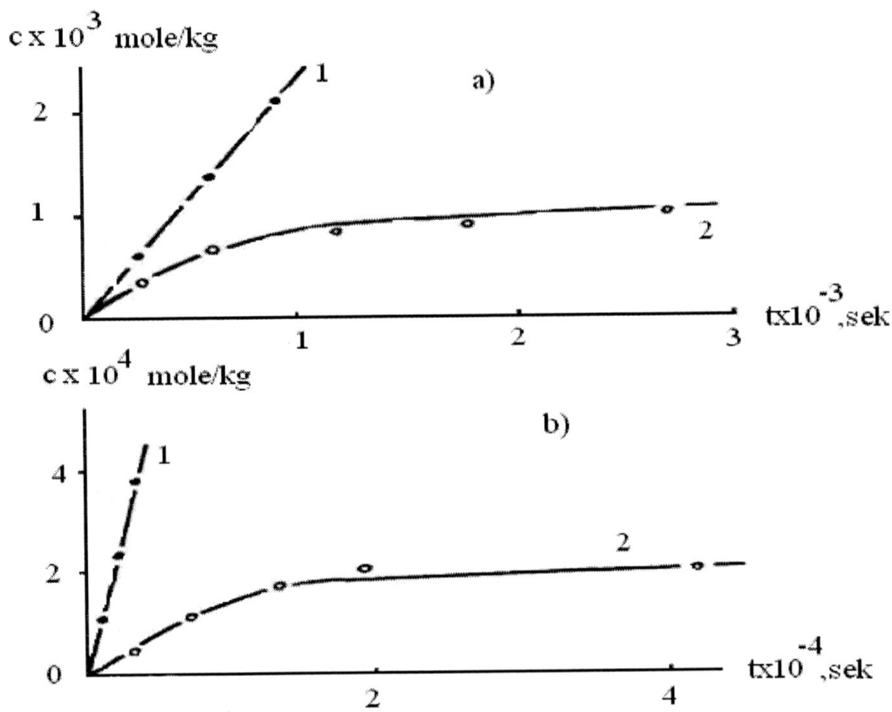

Figure 2.4. Kinetic curves of accumulation of HCl (1) and radicals (2) at CA films irradiation with addition of 0,01 mole/kg of TCMP; light intensity is: a) 90·13, b) 4,4·1013 quant/cm·sec.

This shows that the rate of observed radicals formation is proportional to $I^{1,5}$. So, the reaction of observed radicals formation has the same order, according to light intensity, as the reaction of acetic acid formation. That is why formation of observed radicals and acetic acid proceeds in parallel reactions.

It is interesting to note, that at low light intensities the rate of formation of observed radicals is much lower than W_{in} (Figure 2.4 b). This indicates the presence of high-active radicals in the system with large constants of the destruction rate.

Experiments at lowered temperatures were carried out for discovering these radicals. EPR spectrum, being superposition of singlet and triplet signals (Figure 2.5 a), is registrated at CA irradiation at 20°C. Warming of the samples destroys active radicals. Only stable singlet 1,6 mT wide remains here (Figure 2.5 b). Analogous singlet is observed at TCMP photolysis at 25°C. Difference of the first and second spectra gives EPR spectrum of radicals, destructed at warming up (Figure 2.5 b): it is triplet with STV splitting by 2,9 mT.

Singlet 1,6 mT wide being observed at photolysis and radiolysis of cellulose and its derivatives is attributed to radicals with the chain of conjugation: these are alkyl and polyenyl radicals [148].

Triplet 2,5-3,0 mT wide is attributed to radicals, obtained by breaking of hydrogen atom from carbon in positions 2, 3 and 5: these are hydroxylic radicals in cellulose and acetyl alkyl radicals in cellulose diacetate.

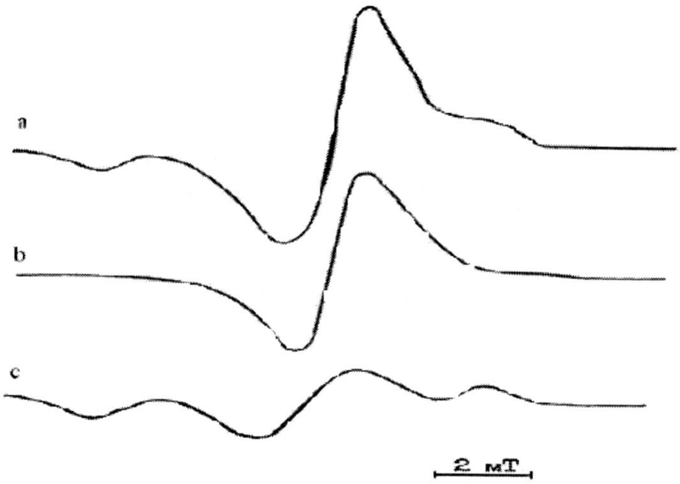

Figure 2.5. EPR spectra at 77°K of CDA-film, irradiated by light (λ=254 nm) in the atmosphere of helium at 253°K for 11 minutes: just after irradiation the same sample after warming up for 3 minutes at 296°K; spectrum, obtained by graphic subtraction of spectrum b) from spectrum a).

Experimental data on the formation of acetic acid and radicals at CA photolysis can be explained by Scheme 3.

$$X + h\nu \rightarrow R^*_1,\ W_{ин} \qquad (0)$$
$$R^*_1 + h\nu \rightarrow R^*,\ K_1 \qquad (1)$$
$$R^*_2 + h\nu \rightarrow R^*_1,\ K_2 \qquad (2)$$
$$R^*_1 + h\nu \xrightarrow{R1H} AcOH + R^*_1,\ K_{3,T} + K_{3,\Phi}\cdot I \qquad (3)$$
$$R^*_1 + h\nu \rightarrow \text{destruction},\ K_4 \qquad (4)$$
$$R^*_1 + R^*_2 \rightarrow \text{destruction},\ K_5 \qquad (5)$$
$$R^*_1 + R_2 \rightarrow \text{destruction},\ K_6 \qquad (6)$$

Scheme 3.

Here: X – chromophore (CA or photoinitiator); R_1* - acetoxyalkyl radical with high reactivity; R_2* - low-active polyenyl radical with free valency, conjugated with double bond; AcOH – acetic acid; K_1 – K_6 – constants of the rate of corresponding reactions. $K_3 = K_{3,T} + K_{3,\Phi}$, where $K_{3,T}$ and $K_{3,\Phi}$ – constants of the rate of the process in the darkness and under light action.

Scheme 3 describes the same principle of radicals photolysis as Scheme 2, and differs from Scheme 2 only by the fact that it includes concrete reactions for those radicals, which are experimentally identified.

By the method of numerical integration of differential kinetic equations of the Scheme 3 it has been found that this scheme quantitatively describes experimental data on kinetics of accumulation of acetic acid and radicals at CA photolysis (Figure 2.1. – 2.4) at the following set of kinetic parameters:

$W_{in} = 2{,}5\cdot10^{-22}\cdot I$ mole/kg·sec; $K_1 = 5{,}5\cdot10^{-17}\cdot I\ sec^{-1}$; $K_2 = 3{,}3\cdot10^{-18}\cdot I\ sec^{-1}$;
$K_3 = 0{,}015 + 5{,}5\cdot10^{-16}\cdot I\ sec^{-1}$; $K_4 = 1000$ kg/mole·sec; $K_5 = 3{,}6$ kg/mole·sec;
$K_6 = 0{,}013$ kg/mole. (where I – light intensity in quantum/cm^2).

The value of K_6 = 0,013 kg/mole·sec has been derived from the data on kinetics of dark destruction of radicals R, having singlet EPR spectrum. The value of other constants $K_1 - K_5$ should be considered as examples, as their real values may be obtained only from measurements of the rates of corresponding elementary stages.

Theoretical curves of AcOH, accumulation (solid lines 1 and 3 in Figure 2.1); dependence of the rate of AcOH formation on light intensity (solid line in Figure 2.2); hysteresis of the dependence of CA photolysis rate on light intensity (solid lines 1 and 2 in Figure 2.3); kinetic curves of formation and accumulation of radicals (solid lines 1 and 2 in Figure 2.4).

Values of the constant, found by selection, are quite possible. Light constants of the rate according to the value order agree with theoretical values of constants on the basis of quantum yields (Φ = 0,1 – 1) [167], being characteristic for radicals, and extinction coefficients ($\varepsilon - 10^4$l/m·cm). Large value K_4 = 1000 kg/mole for acetyl alkyl radicals may be caused by acetic acid, taking part in integration of free valency. It is known [168], that small quantities of low-molecular substance increase the rate of free valency travel by 3-7 orders in comparison with the rate of usual chemical reaction.

Formation of acetic acid in the darkness is explained by radical-induced deacetylation according to total reaction

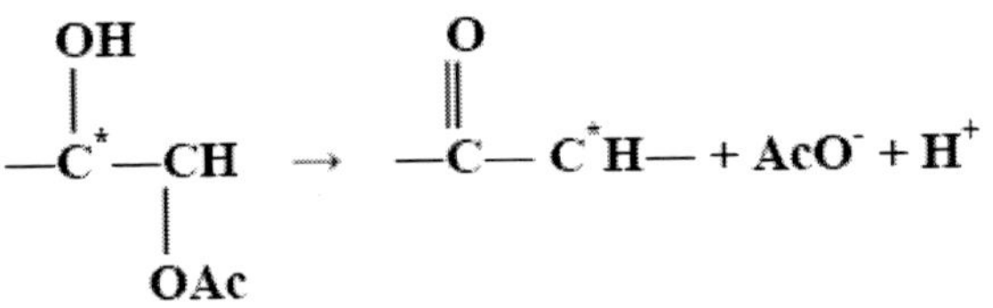

(1)

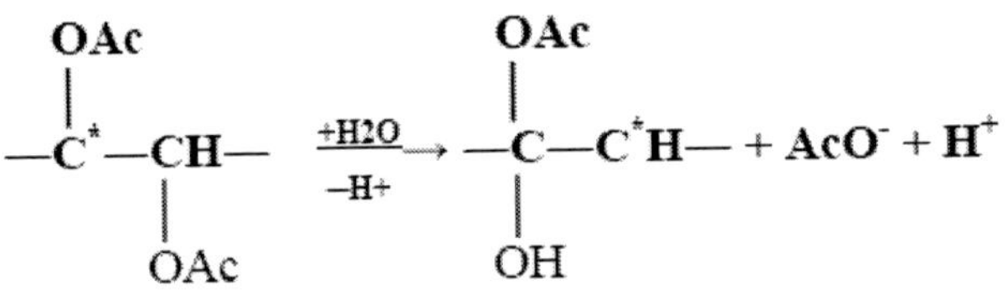

(2)

Radical=induced deacetylation proceeds in the darkness in several stages according to the mechanisms S_{N1} or S_{E2} in the darkness [169].

Found low value of activation energy of the process of AcOH formation in the darkness (E_T=42-54 kJ/mole) is characteristic for the reaction of radicals decay and agrees with the given mechanism.

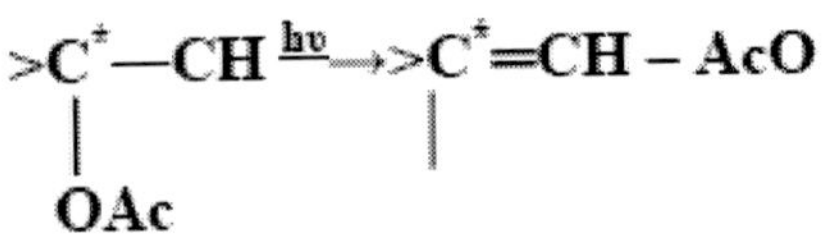

(3)

As it is seen from Scheme 3, central reaction of CA photolysis is photoreaction (3) of acetoxyalkyl radical R_1*, presenting chain photoprocess of acetic acid formation. So, it is necessary to discuss the ideas on mechanism of acetic acid formation, being presented in literature.

Before it has been supposed that AcOH is formed from acetoxy-radical AcO* as a result of its splitting off valency-saturated CA molecule under the light action (O). Data show that this is not so. Besides, it is stated [170] that AcO* radicals do not have end life-time and instantly decay with carbon dioxide splitting off according to the reaction

$$AcO^{*} \rightarrow CH_3 + CO_2 \quad (4)$$

In connection with this, acetic acid is not formed from acetoxy-radicals and this allows to exclude them from the reaction (3).

Another possible photoreaction of acetoxyalkyl radical – is acetyl radical Ac* splitting off.

$$C^{*}\text{-}OAc + h\nu \rightarrow C\text{-}O + Ac^{*} \quad (5)$$

Transition of EPR triplet spectrum with splitting of 2,8 mTl into singlet spectrum 0,9 mTl wide at polyvinylacetate [171] is explained by photoreaction (5). However, acetyl radical in vacuum does not lead to formation of acetic acid. That is why the reaction (5) does not explain large yields of AcOH at photolysis of polyvinylacetate and CA and, probably, it does not present the main way of phototransformation of radicals R*.

At the same time photoreaction of acetyl alkyl radicals (reaction 3 in scheme 3), leading to chain process of acetic acid formation, indicates the proceeding of Norrish reaction of the 2nd type,

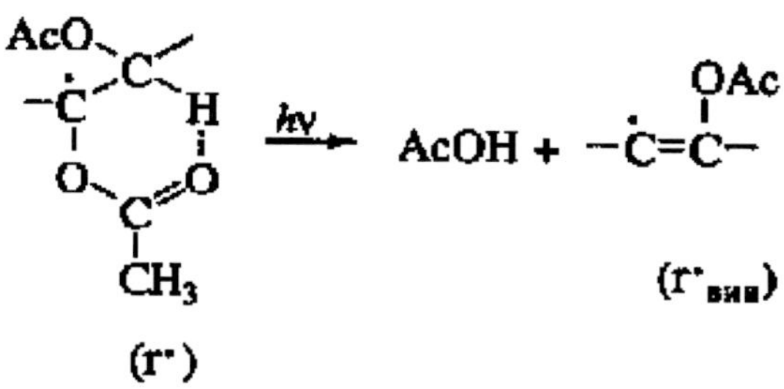

(6)

Norrish reaction of the 2nd type is well known for ketones and esters. Acetyl group in radical R has greater degree of freedom than in valency-saturated molecule RH, and this facilitates the formation of intermediate six-member complex, through which Norrish reaction proceeds.

Vinyl radical R_{vin}*, being formed during reaction (6) simultaneously with AcOH, is very active even at lowered temperatures. For example, they easily break chlorine atom from CCl_4 [160]. That is why vinyl radical, being formed during reaction (6), easily breaks hydrogen atom from polymer, giving again R according to the reaction

$$\underset{\displaystyle -\overset{|}{C^{*}}=CH-}{OAc} + RH \rightarrow \underset{\displaystyle -\overset{|}{C}H=CH-}{OAc} + R^{*} \quad (7)$$

Thus, sequence of reactions (6) and (7) provides chain process of acetic acid formation (see reaction 3 in Scheme 3).

Formation of polyenes in polyvinylacetate [166] and formation of double bonds in CDA [162] may be now explained by this sequence of reactions. In the case when RH has polyene fragment $R_{polyene}H$, reaction (7) will lead to the formation of low-active polyene radical.

$$-\overset{\overset{AcO}{|}}{C^*}=C- \; + R_{polyene}H \rightarrow \overset{\overset{AcO}{|}}{C}H=C- + R^* \qquad (8)$$

Reaction (6) of photolysis, followed by fast reaction of polyene radical formation (8), is unfolded 1st stage in the Scheme 3. Obtained values K_1 and K_3 (K_3 being larger than K_1 by 10 times) show that reaction (8) is not the main reaction of vinyl radicals. It can be explained by low concentration of double bonds.

Thus, reactions (6) – (8) explain chemical mechanism of acetic acid chain formation at CA photolysis and agree with kinetic scheme of the process. According to the scheme, length of the chain of acetic acid formation depends on light intensity and at intensities, being used, it is 10-30 units.

2.4. Kinetic Regularities of CDA Photooxidation

Kinetic curves of O_2 absorption at CDA irradiation by light (253,7 nm) are given in Figure 2.6. As it is seen from Figure 2.6 curves of O_2 absorption consist of two sections: first-effect of photochemical preaction, characteristic for cellulose materials, is observed at initial stages of photooxidation at photooxidation levels lower than $2{\cdot}10^{-5}$mole/kg; second – stationary, from the slope of which the rate of oxygen absorption is found in stationary mode. As it is seen from Figure 2.6, oxygen absorption proceeds with the rate a bit higher zero from the very beginning, however, at accumulation of radicals and achievement of stationary concentration, the rate of oxygen absorption increases by 2,1 times.

Rise of the rate of oxygen absorption at initial stages is caused by increase of radicals concentration up to its stationary value that is in conformity with increase of (R*) reaction rate increases [172].

$$R^* + O_2 \rightarrow R^*O_2$$

The fact, that at the very beginning of irradiation the rate of oxygen absorption is not equal to zero, indicates the possibility of high-active radicals taking place in CDA photooxidation. On the other hand, the time of achieving stationary rate of oxygen absorption according to data of Figure 2.6 is 300 seconds, that shows fast achievement of stationary radicals concentration.

As it is seen from Figure 2.6 (curve 2) at the moment of light source switching off the rate of O_2 absorption decreases up to stationary rate in dark process and oxidation reaction proceeds for some time. This effect is the effect of photochemical aftereffect or post-effect. It is explained [173] by lasting radical-chain process of oxidation, consisting of two stages: reactions of continuation and breaks of chain of photooxidation. Kinetic curve of oxygen absorption in post-effect at quadratic break of oxidation chains in the presence of dark process of initiation is described by the equation [174]:

$$[\Delta O_2]=Kn[RH]/Kr*\ln((1+\xi bh(t/\tau st)/1+bh(t/\tau st))+Wst*t \quad (1)$$

where, K_n and K_r – constants of rates of reactions of continuation and break of oxidation chains, [RH] – concentration of reactive part of monomeric units, $\tau_{st}=1/K_r[R*O_2]_{st}$ – life-time of radicals in stationary mode of oxidation, $[R*O_2]_{st}$ – concentration of radicals in stationary mode, $\xi=W_0/W_{st}$ (where, W_0 and W_{st} – rates of oxidation at the initial moment of time after switching off the light and in stationary mode of dark oxidation, correspondingly.

Provided, that t » t_{st}, equation (1) assumes the form:

$$[\Delta O_2]=Kn[RH]/Kr*\ln((1+\xi)/2)+W\text{ст}*t \quad (2)$$

According to the equation (2), experimental data in $[\Delta O_2]$ – t plot of large t is described by strait line, tangent of slope angle of which is equal to W_{st}, and cutting on Y-axis is equal to

$$[\Delta O_2]=Kn[RH]/Kr*\ln((1+\xi)/2) \quad (3)$$

Substituting experimental value of ξ=5,3 into the equation (3) important parameter, characterizing chain photooxidation has been found K_n [RH]/K_r = 0,9·10^{-5} mole/kg.

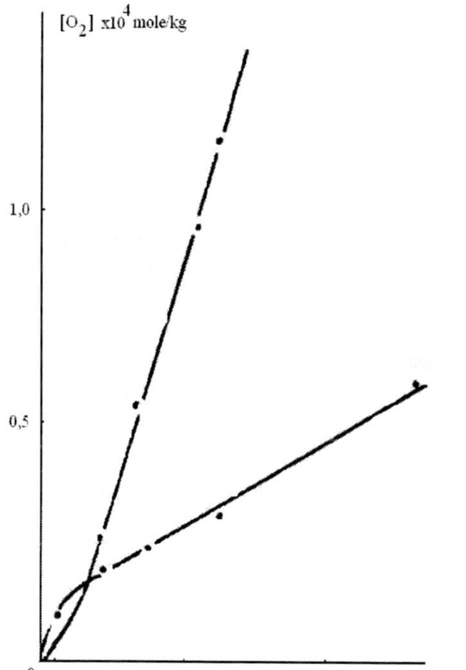

Figure 2.6. Kinetic curves of oxygen absorption at irradiation (1) and after irradiation (2) of CDA by the light with wave length of 253,7 nm and intensity 1-0,5·10 quant/cm"#s at 298°K and pxygen pressure 150 T. 2-theoretical curve of photochemical aftereffect, calculated according to the equation 3.1 at Kn[RH]/Kr=o,9-10"5 mole/kg, y=5,3; T=300s; WTeM=1,69-10"8 mole/kg·s.

As it is seen from Figure 2.6, experimental data according to post-effect are well described by the equation (1) at the values: ξ=5,3; $K_n[RH]/K_r$= $0{,}9 \cdot 10^{-5}$ mole/kg; τ_{st}=300c and W_{sg}=1,69·10 mole/kg·s. Correspondence of obtained data with kinetic law of photochemical aftereffect at quadratic break of oxidation chains with equation (1) shows that chains break at CDA oxidation is quadratic.

2.5. Light Stabilization of CDA by Hexaazocyclanes

It has been noted that CDA photolysis proceeds with participation of two active centres, order 1,5 according to light intensity for the rates of formation of acetic acid and polyene radicals, that gives the possibility to determine mechanism of CC action.

Since inhibitor of radical processes are low-effective light stabilizers of CA, then it is necessary to search for compounds with large values of extinction coefficient as CA light stabilizers. For this purpose we have used hexaazocyclanes (HAC), which have ε=1-$10 \cdot 10^4$l/mole·s over the range of λ=200-350 nm.

Hexaazocyclanes (HAC), having unfolded conjugated chain, high photostability and thermal stability, were investigated as CC, and presence of chromophoric groups assumed to use them as dyes for polymers.

Chemical structures of chosen HAC differ in degree of conjugation and flatness. For example, HAC on the basis of phthalodinitrile and n-phenyldiamine (HAC – XLIX) – is conjugated nonflat system [167]. HAC on the basis of phthalodinitrile and 9,9 – bis (4 diamine) phlourene (HAC – LII) is nonconjugated cycle.

We have also investigated substances (HAC – LVII), (HAC – LI), being the products of condensation of phthalodinitrile with para-phenylene diamine in order to examine the effect of macrocyclic stabilizer structure on HAC properties.

Light-stabilizers, introduced into polymer, in general case may simultaneously act according to several mechanisms: ultra-violet shielding, inhibition; and in some cases display undesirable effect of photosensibilization.

Taking into account coefficient i_φ, efficiency of stabilizer A was defined according to the following equation

$$A = W_0 / (W_{st} \cdot i_\varphi) \qquad (4)$$

If the additive acts only as ultraviolet shield then A=1. In the case when the additive acts as sensitizer, then A<1. And if A>1 then effect of overshielding.

Value A for all investigated light stabilizers is less than 1. Hence it follows that HAC acts not only as ultraviolet shields, but as photosensitizer too. It is seen from Table 13A that the best light stabilizers are HAC-1, P-3 and HAC-4, for which value A is the highest. Introduction of HAC inhibits the process of photomechanical CDA destruction by 5,1-5,9 times.

It should be noted that HAC-1 does not prevent oxidation after CDA irradiation. Taking into account coefficient of shielding in the case of stabilized CDA films it has been found, that kinetics of O_2 absorption after irradiation is described by equation (1) at $K_n[RH]/K_r$=$0{,}9 \cdot 10^{-5}$ mole/kg and the rest of parameters are the same, as in the case of initial

CDA. Proceeding from these two factors one may come to a conclusion that HAC does not act as an inhibitor.

Table 13A. Dependence of stabilizer efficiency on its structure (concentration of stabilizer is 5% of polymer mass)

Stabilizer	$A=W_0/W_{st}\cdot i_\varphi$	Stabilizer	$A=W_0/W_{st}\cdot i_\varphi$
HAC-1	0,111	HAC-6	0,072
HAC-2	0,150	HAC-7	0,070
P-3	0,230	HAC-8	0,084
HAC-4	0,260	P-9	0,111
HAC-5	0,111		

If the additive does not act as inhibitor, but only as ultraviolet shield and photosensitizer, then dependence of quantum yield of photooxidation ΦO_2 on light intensity I_0 at full absorption of incident light should be described by well-known equation (5)

$$\Phi o_2 = W/I_0 + \alpha\Phi_i + K_n[RH]/\sqrt{K_r}\cdot\sqrt{\Phi}\cdot 10^{-3}\cdot 1/\sqrt{I\varepsilon C} \quad (5)$$

where: W – the rate of O_2 absorption in terms of $1 cm^2$ of the sample surface, mole/$cm^2\cdot s$; Φ_1 – quantum yield of photoinitiation, caused by photosensitization of the dye; α – quantity of O_2 molecules absorbed in non-chain process of oxidation per one formed radical; ε – extinction coefficient; C – concentration of the additive.

Large photosensitizing activity of HAC – LI is successfully explained by the fact that compound (LI) may have two tautomeric resonance states, in one of which it contains free aminogroups, owing to which it displays photosensitizing properties, that quite well agrees with [10].

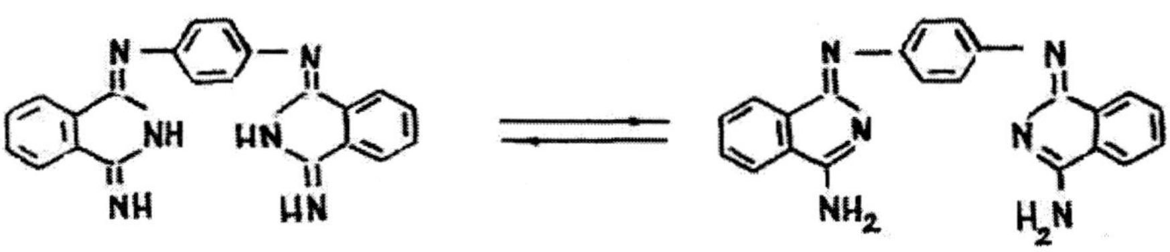

Light stabilizing action of HAC – XLIX and HAC – L is proved by the data on decrease of reduced viscosity of CDA acetone solution after ultraviolet irradiation of films, containing HAC – XLIX and HAC – L (Table 14).

Investigation of CDA photooxidation was carried out in order to get additional information about light-stabilizing action of HAC. Figure 2.7 shows dependence of concentration of absorbed oxygen (ΔO2) on the time of irradiation at photooxidation of three CDA samples, that is of unstabilized film (curve 1) and stabilized HAC – XLIX (curve 2) and 2,2,6,6 – tetramethyl – 4 – hydrooxipiperidine – 4 – hydroxyl – 1 – oxyl – (mitroxylic radical) (curve 3). As it is seen from Figure2.7 HAC decreases the rate of CDA photooxidation.

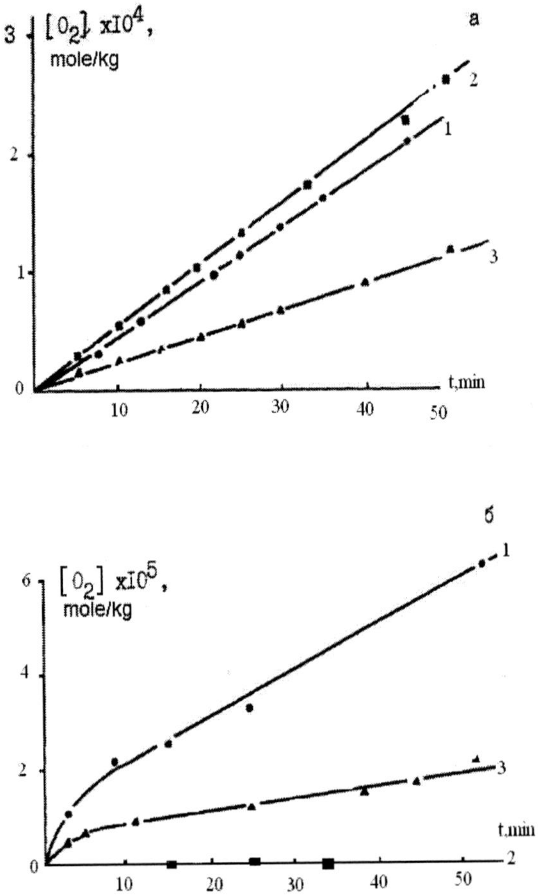

Figure 2.7. Kinetics of O2 absorption at irradiation (a) and after irradiation (b) of CDA without additive (1) and with 0,01 m/kg (2) and 0,016 m/kg of HAC – XLIX (3).

Table 14.

Stabilizer	Concentration, %	Concentration of reduced viscosity ofсти 0,5% CDA solution after irradiation in %	
		τ=24 hours	τ=48 hours
-	0,0	13	7,0
	0,5	25	-
TCK	2,0	79	-
	5,0	70	-
HAC-XLIX	0,5	79	-
	2,0	92	-
	3,0	94	66,0
	5,0	93	-
HAC-L	0,5	-	31,0
	2,0	-	74,0
	5,0	-	90,7

On the other hand, inhibiting effect of nitroxylic radical at irradiation is not observed owing to its strong action as photosensitizer. Photosensitization is absent after irradiation in dark reaction and nitroxylic radical displays its inhibiting action. That is why oxidation after irradiation (post-effect) is not observed (Figure 2.7, curve 2).

It should be noted that HAC – XLIX does not suppress CDA oxidation in spite of the fact that concentration of HAC – XLIX is larger than that of nitroxylic radical by 1,6 times.

Equation (1) was used to define kinetic parameters of oxidation. It has been found that the value K_nRH/K_o=0,9·10^{-5}mole/kg for CDA in the presence of HAC – XLIX is close to the value K_nRH/K_r=1,0·10^{-5}mole/kg for unstabilized CDA.

So, we may come to a conclusion that HAC – XLIX is not an inhibitor of radical-chain photooxidation of CDA.

If the additive is not an inhibitor and acts only as ultraviolet absorber and sensitizer, then dependence of quantum yield of photooxidation on light intensity at strong light absorption [168] should satisfy the equation (6):

$$\Phi o_2 = \alpha\Phi_i + K_n[RH]/K_0 \cdot\sqrt{\Phi_i}\cdot 10^{-3}\cdot 1/\sqrt{I_0\varepsilon C} \qquad (6)$$

where: Φ_i – quantum yield of photoinitiation of the dye caused by photosensitization; α – quantity of oxygen molecules absorbed during non-chain photooxidation per 1 formed free radical; $\varepsilon_1 C$ – extinction coefficient and dye concentration.

Figure 2.8 shows that experimental dependence of Φo_2 on light intensity I_o at CDA photooxidation in the presence of HAC – XLIX agrees with theoretical ones according to the equation (6).

Calculations according to the equation (6) show that the value Φ_i achieves 0,004. Hence, it follows that HAC – XLIX, introduced into CDA, is not an inhibitor. It acts as ultraviolet absorber and photosensitizer oxidation according to radical mechanism.

Obtained result allows to explain dependence of efficiency on concentration of HAC – XLIX (Figure 2.9). Photosensitizing is absent and HAC acts only as ultraviolet absorber at small concentrations of CC (0-1,5%), where the rate of photoinitiation, caused by HAC, is much lower than the rate of photoinitiation, caused by pure CDA. Light stability of polymer increases by i_φ times over this range of concentration A=1. Photosensitizing becomes visible (A<1) at relatively higher concentration of HAC.

As it follows from above-mentioned conclusions, decrease of photosensitizing action of HAC has to improve its light-protective effect.

All investigated compounds have C=N group. Photochemical reactions of C=N – group are similar to carbonyl group [178]. That is why breaking of hydrogen atom from substrate by excited state of C=N – group should be primary photochemical reaction of stabilizer. In the case of HAC – LVI and HAC – LVIII additional way of sensibilization, caused by rhodamine fragment, introduced into HAC structure, may take place. But value A for HAC – LVI (A=0,072) and HAC – LVIII (A=0,111) is of the same order, as for other HAC. Hence, rhodamine fragment does not probably influence photosensitization. It should be noted that intramolecular transfer of hydroxyl proton is possible in excited state of HAC – LVIII through six-membered cycle.

Intramolecular transfer of proton suppresses reaction of photoinitiation [10] and should lead to the greatest value of A for HAC – LVIII. Experimental data do not agree with this one, as value A=0,111 for HAC – LVIII is less, than value A=0,26 for HAC – L.

Thus, the rate of deactivation of excited states of all investigated HAC – stabilizers is much larger, than the rate of intramolecular transfer of proton. Large difference is observed between values of A for HAC – XLIX and HAC – L in spite of small difference in their chemical structures. Computer calculation of electronic structure of HAC – XLIX and HAC – L (Figure 2.10-2.12) shows that HAC – L structure is more planar than that of HAC – XLIX.

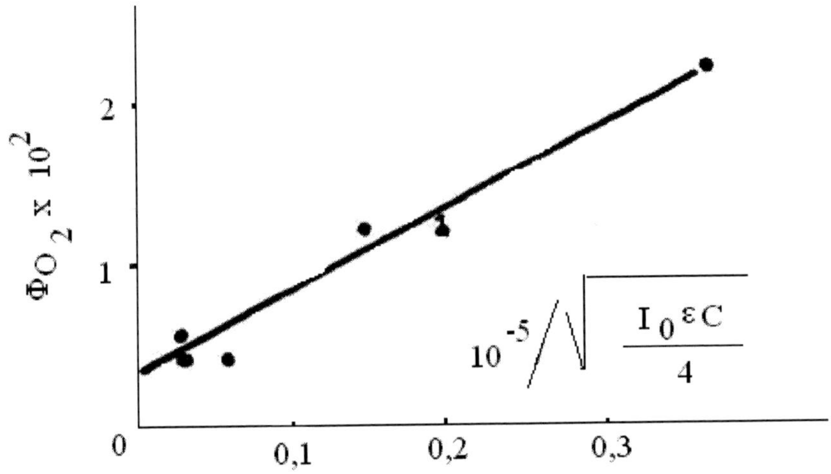

Figure 2.8. Dependence of quantum yield of O2 on light intensity in CDA film + HAC – XLIX.

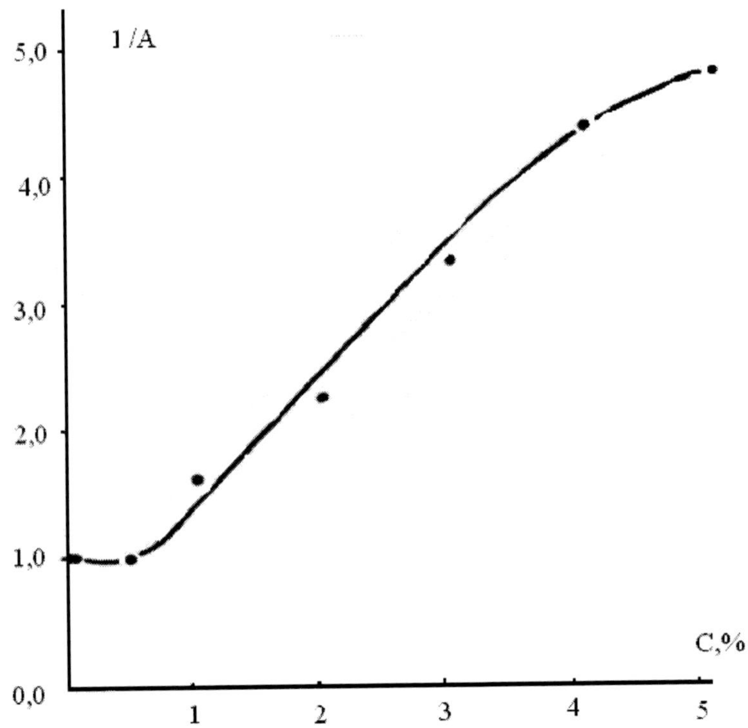

Figure 2.9. Dependence of HAC – XLIX efficiency on its concentration.

Figure 2.10. Structural model of HAC – XLIX.

Figure 2.11. Structural model of HAC – L.

Figure 2.12. Structural model of HAC – LII.

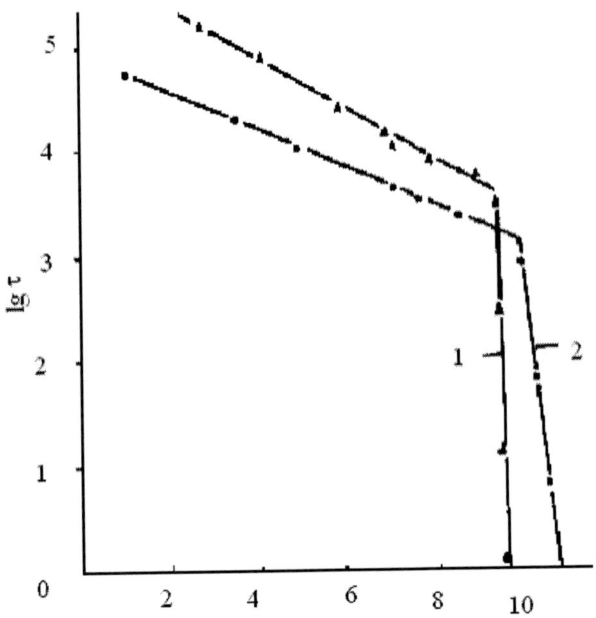

Figure 2.13. Dependence of durability of initial (1) and dyed (2) CDA samples on the load value.

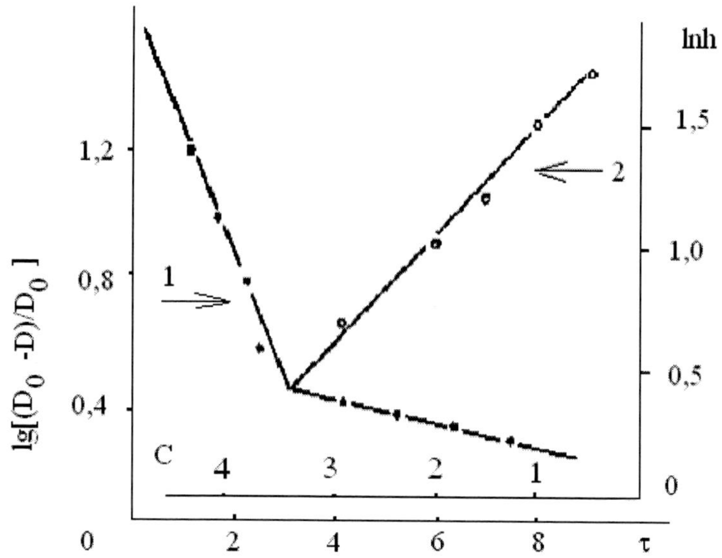

Figure 2.14. Kinetics of Isoramnetine consumption (1) and change of its rate at ultra-violet irradiation (2).

Plane structure of HAC – L should facilitate aggregation of CC molecules, that leads to very effective suppression of excited states. Obtained results give the reason to believe that search for effective photostabilizers should be carried out among macrocyclic aromatic compounds, having structure approaching to a plane.

Obtained effect of stabilization by hexaazocyclanes appeared to be effective and at simultaneous action of mechanical load and ultraviolet light. As dependence of durability of initial and dyed samples of CDA on the value of the load (Figure 2.13) at photomechanical destruction shows, the durability of stabilized sample is higher than initial one by 1,3 times.

Important property of hexaazocyclanes is their ability to dye acetates into yellow, red, brown coloures, since in this case additional introduction of dye is not required.

Method of determination of the mechanism of CC action was successfully used while studying light-stabilizing action of isoramnetine. Isoramnetine (3, 5, 7, 4 – tetraoxy – 3 – methoxyflavone) LXV, being industrial waste of solvent cake of fruits of sea-buckthorn, is perspective as the dye of natural origin for products from cellulose acetate. Absorbing in region λ=256nm close to CDA absorption, isoramnetine also has phonolic hydroxyls and quinoid group in its structure, that supposes its high light-protective activity. It has been found that lightfastness of CDA increases by 2,8 times at isoramnetine concentration equal to 3 mass %; polymer is being dyed steadily along the whole depth and intensity of painting is proportional to the additive content. Changing of painting owing to additive consumption (fading) is being observed during ultraviolet irradiation of CDA film, containing isoramnetine. Constant of isoramnetine consumption (Figure 2.14, curve 1) has been defined at initial section of kinetic curve of the change of optical density value. As it is seen from Figure 2.13 (curve 2), the rate of additive consumption is inversely proportional to its concentration. These results well agree with generally accepted information that at low concentration molecules of the additive equalize among polymer chains forming monolayer [175]. During increasing of additive concentration interaction forces between them overcome interaction forces between additive and polymer, in our case – between isoramnetine and

CDA, and, at last, the whole additive passes into aggregates, which fade slower owing to the effect of concentration suppression.

2.6. Light Stabilization of CDA by Polyconjugated Azomethine Compounds

Use of low-molecular stabilizers has essential shortage – protective effect duration time under working conditions, as they are quickly consumed at ultraviolet irradiation, washed out from the polymer and they are lightly volatile products.

In this connection there rises a question about use of highmolecular light-stabilizers. However, very little attention is paid to these stabilizers in literature. Besides, there are not enough systematic investigations of mechanism of the action of polymer light-stabilizers in literature. Polyconjugated azomethine compounds (PAC) attract attention from the set of potential highmolecular light-stabilizers; stabilizing activity of PAC regarding CDA has been found by I.Ja. Kalontarov with his helpers [176].

Interesting results have been obtained during investigation of lightfastness of products from cellulose diacetate by the way of using additives of polyconjugated azomethine compounds, their characteristics are presented in Table 15, and studying the mechanism of these compounds effect on CDA transformations on photooxidative destruction. Absorption ability of PAC-products at ultraviolet irradiation is in the range of 253 – 368 nm, that supposes presence of shielding mechanism of polymer protection at introducing of these chemicals – additives.

Measurement of acetic acid accumulation at ultraviolet irradiation of stabilized and unstabilized CDA film were carried out for quantitative evaluation of stabilizing activity of PAC as ultraviolet absorbers. Relation of the rates of acetic acid in the absence (W_o) and presence (W_{st}) of additives was recognized as characteristic of phototransformation [10].

Effect of well-known industrial ultraviolet absorbers of Tinuvine and 2-oxy-4-methoxybenzophenone type, that were not inhibitors of radical processes, was investigated for comparison [162]. Being measured values of the rate of acetic acid accumulation were obtained with regard to corrections for shielding.

Table 15. Characteristics and light-stabilizing properties of PAC

Stabilizer	Average molecular mass	T_{melt} ^{0}C	λ, nm in C_2H_5OH	Conservation in % spec.
XXXI	400	140-145	261;360	81,0
XXXII	800	180-186	266;350	85,0
XXXIII	1400	116-120	352	82,5
XXXIV	600	90-95	253;372	
XXXV	1000	110-115	263;368	84,0
XXXVI	1200	210-215	258;372	77,0
XXXVII	1680	103-105	277;315	

From Table 16 it follows that tinuvine II and 2-4-methoxybenzophenone decrease the rate of CTA phototransformation at rather large concentrations, while, as in the case XXXV, the same effect is achieved at relatively small concentration. Since for XXXV- $W_o/W_{st} \cdot i_\varphi > 1$, then, in the given case, there is additional mechanism of protection, excepting shielding; specifically XXXV may play the role of antioxidant, inhibitor of radical processes or suppressor of excited states.

Table 16. Effect of additives of ultraviolet absorbers on the rate of acetic acid accumulation at irradiation of CDA in the air by the light (253,7 nm)

Additive	% of CA mass	i_φ	W_0/W_{st}	$W_0/W_{st} \cdot i_\varphi$
Tinuvine II	2,0	1,66	2,10	1,30
Tinuvine II	3,8	1,71	3,50	2,00
2-oxy-4-methoxybenzophenone	2,0	1,50	1,40	0,90
2-oxy-4-methoxybenzophenone	3,8	1,61	2,10	1,30
XXXV	0,5	1,22	2,00	1,64
XXXV	2,0	2,55	4,38	1,72
XXXV	3,0	5,20	7,78	1,50
XXXV	3,8	3,77	4,05	1,07

CA irradiation of stabilized XXXV in the air and in vacuum under the same conditions was carried out to prove antioxygenic activity of PAC, taking the rate of acetic acid accumulation for characteristic of phototransformation with regard to correction for shielding. Data of the experiment are given in Figure 2.15 (curves 2 and 3).

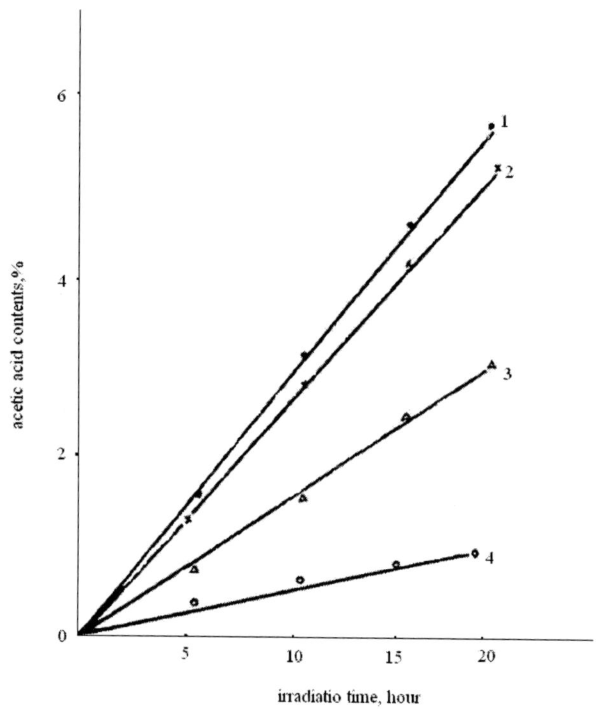

Figure 2.15. Content of free CH3COOH in the process of irradiation be the lamp PRK – 2. 1 – CDA – film without additive; 2 – CDA – film, containing BA; 3 - CDA – film, containing XXXIV; 4 - CDA – film, containing XXXV.

It is seen from this figure that data, obtained as a result of these experiments, are identical, that proves antioxygenic action of XXXV.

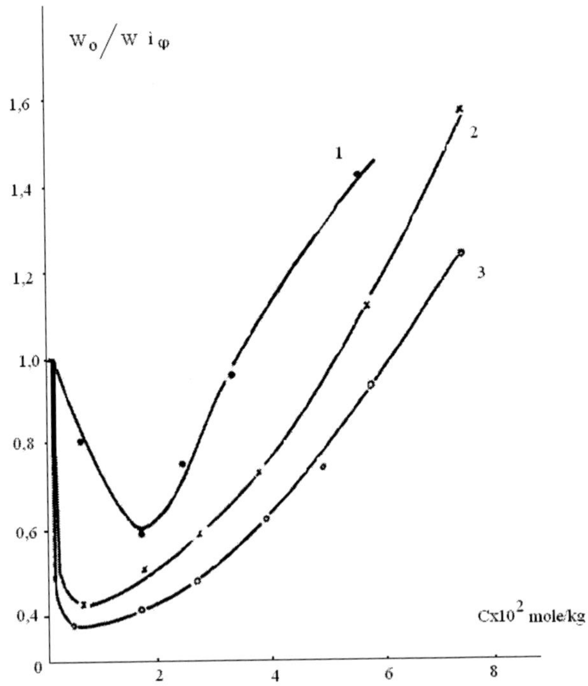

Figure 2.16. Dependence of the rate of acetic acid accumulation on additives concentration: 1 – Tinuvine in the air; 2 – XXXV – in the air; 3 – XXXV in vacuum.

Comparisons of XXXV with well-known antioxidant – Tinuvine over the wide range of concentrations have been carried out in further experiments. Data of these experiments are given in Figure 2.16, from which it is seen that XXXV, similar to Tinuvine, accelerates phototransformation in the range of large concentrations, and in the region of small concentrations it decelerates this process. Low value of maximum concentration of XXXV may be connected with the fact that reaction proceeds in noncrystal regions of polymer, which are more available for additives. It should be noted here that XXXV acts as antioxidant at much lower concentrations than Tinuvine.

The fact that XXXV displays light-protective activity, decelerating CDA destruction at ultraviolet irradiation not only in vacuum (Figure 2. 16, curve 3), but in the air (Figure 2.16, curve 2), moreover, curves 2 and 3 have the character close to curve 1, that undoubtedly testifies in favour of XXXV action not only as ultraviolet absorber, but also as an antioxidant. Possibility to decelerate the rate of CDA photodestruction by the products of inhibitors transformation is important, too.

It is known that PAC plays the role of inhibitors of radical-chain processes, specifically, polymerization [177]. In all probability, interaction of PAC molecule with being formed radicals in CDA at irradiation takes place. Reaction between radicals and inhibitor with formation of completely inactive products may probably proceed here:

$$P^* + I_n \xrightarrow{K1} \text{inert products}$$

Expected stationary state may be good approximation for calculation of kinetics of reaction flow. Mechanism of inhibition by azomethine conjugated compounds is accompanied by the break of multiple carbon-nitrogen bond and addition of radical. In the case of low-molecular azomethines separation of hydrogen atom and addition of radical to α – carbon atom take place:

$$P^{*} + C{=}N \rightarrow PH + C\text{-}N^{*}\text{-}$$

Disappearance of conjugated bonds in the process of radical reaction was proved by data of infrared-spectra. Inhibition by end primary NH_2 – groups proceeds with hydrogen atom breaking off:

$$P^{*} + RNH_2 \rightarrow PH + RN^{*}H$$

However, one should take into consideration that fact, that atomic radicals, being formed, especially of tertiary atom, may take part in the chain transmission onto monomer [178]. Kinetic dependences of the rate of inhibited radical polymerization of styrene in the presence of XXXV, XXXVI, XXXVII (Figure 2.17 – 2.19) show that the rate of polymerization reaction decreases with the increase of azomethines concentration and depends on structure, structural features and efficiency of the chain of PAC conjugation.

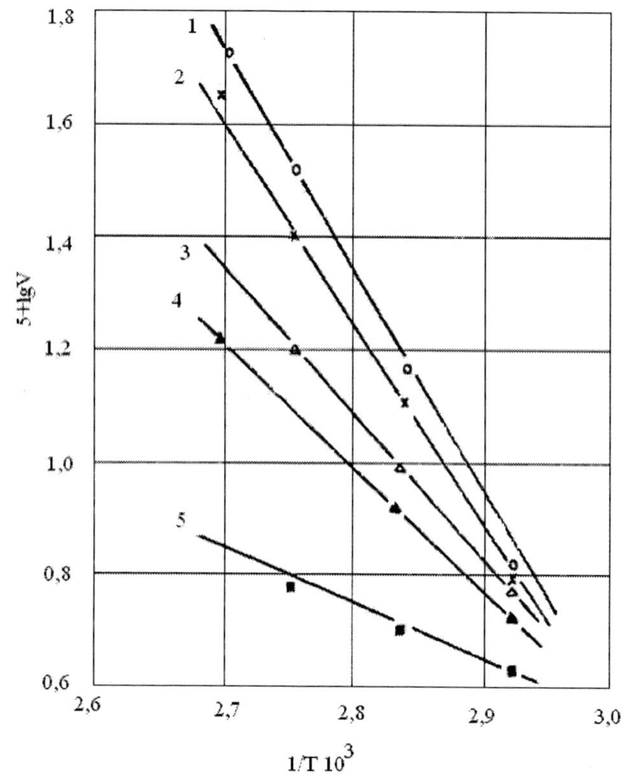

Figure 2.17. Dependence of /V/ styrene on temperature and structure of oligomer: 1 – without additives; 2 – 5 – with the addition of 0,0005 mole/l oligomer: 2 – XXXIII 3 – XXXV 4 – XXXVI 5 – XXXVII.

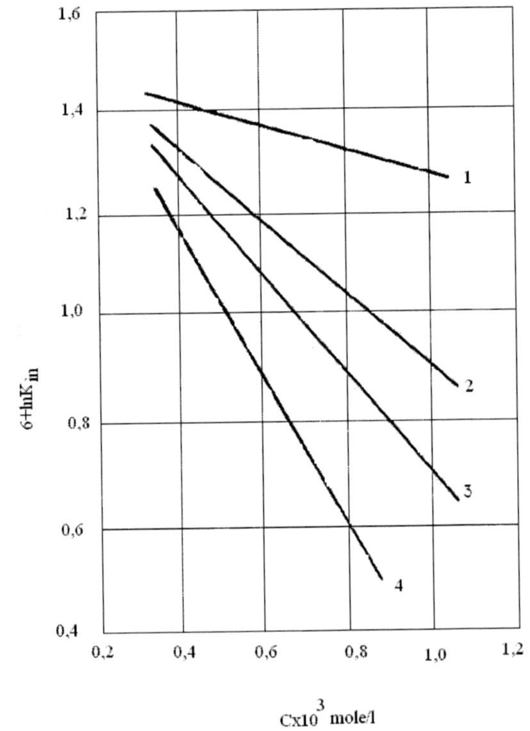

Figure 2.18. Effect of oligomeric system concentration on the constant of polymerization rate at 80°C 1 – XXXIII 2 – XXXV 3 – XXXVI 4 – XXXVII.

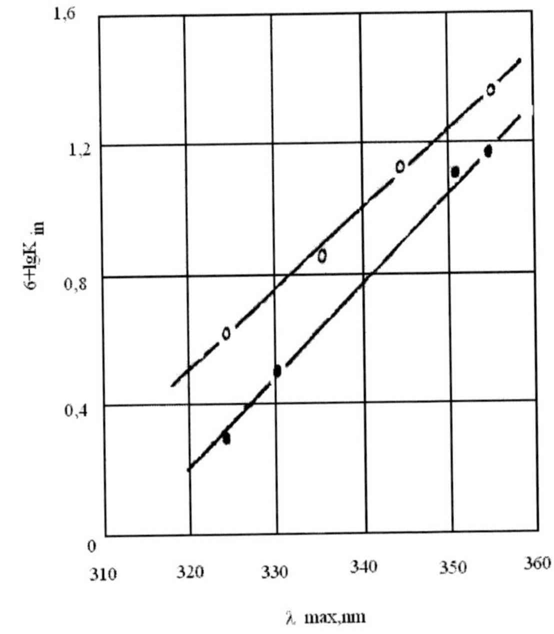

Figure 2.19. Dependence of polymerization rate on the degree of azomethines conjugation at 80°C 1 – low molecular azomethine BA/c/=0,001 mole/l 2 – oligomer XXXV/c/ - 0,001 mole/l.

Hence, one may come to a conclusion that oligomeric azomethine additives are, as a matter of fact, regulators of radical polymerization of styrene and may be also used and for this purpose. PAC products contain undivided ρ – electrons of heteroatom of nitrogen in the chain of conjugation. That is why energy of conjugation (delocalization of PAC electrons) is defined by the presence of not only π – π conjugation, when efficiency of delocalization is $\varphi=\Delta E/\Delta N \cdot \pi$, that is, it is defined by the energy of π – electrons transition from the basic to excited state and depends not only on the length of the conjugation chain, but on competing effect of ρ- π conjugation [179]. Owing to this, disproportionation of electronic density takes place, the value of effective conjugation decreases and, unlike polyenes, bathochronic shift with the growth of the length of conjugation chain is not observed.

It may be noted that efficiency of oligomeric PAC is more than 10 times higher comparing with low-molecular analogs (Figure 2.19). Hence it follows that the increase of conjugation efficiency decreases reactivity of oligomers in the range – XXXVII>XXXV>XXXIII.

Inhibitive activity grows with the rise of temperature of radical reaction, that is characteristic for polyconjugated systems.

It is known that reaction activity of inhibitors in radical reactions may extremely change depending on the inhibitor concentration.

Carried out investigations have shown (Figure 2.18) that the rate of polymerization depends on the structure of being introduced oligomeric inhibitors and increases with the growth of its concentration from 0,0003 to 0,001 mole/l.

So, there has been drawn a conclusion, that oligomers with conjugated azomethine bonds are more effective, as inhibitors of radical reactions, than low-molecular analogs. Their reactivity depends on the length of conjugation chain, its efficiency and on the concentration of oligomer being introduced.

On the basis of performed investigations one may come to a conclusion that mechanism of PAC stabilizing action is not unique and is caused by totality of effects of shielding, inhibition of radical processes and acting as antioxidant.

Change of specific and intrinsic viscosity of CA, destruction of ester groups, accumulation of acetic acid and carbonyl groups are the result of photochemical transformations taking place in CA under the light action.

Above-mentioned facts, caused by CA phototransformations during irradiation by different light sources, have been investigated, using methods of chemical analysis, viscosimetry and spectroscopy, since it is known that phototransformation during irradiation by different light sources, may differ in quantitative and qualitative ratio. Benzalaniline (BA), XXXIV, XXXV, XXXVI, XXXI, XXXII have been used as additives, BA being model compound.

Lightfastness of CA with and without the additive is characterized by the change of specific and intrinsic viscosity of CA before and after irradiation by mercury-quartz lamp PRK – 2. Data on PAC additives effect (2% from polymer mass) are given in Table 17.

From these data it is seen that introduction of PAC into CA slightly changes indices of polymer viscosity in comparison with initial CA before irradiation. Data of Table 6 show that azomethines XXXIV, XXXV, XXXI, XXXVI, XXXVII content in CA decreases specific viscosity fall.

It also follows from Table 6 that introduction of 2% of PAC from CA mass facilitates considerable conservation of initial indices of CA at irradiation. So, intrinsic viscosity of CA solution, not containing light-stabilizer, after 10 hours of irradiation decreases by 68,6%, but in films, containing XXXV, XXXI, XXXII, XXXVI, XXXVII – it decreases only by 13-15%. These PCA appeared to be the most effective light-stabilizers.

Table 17. Change of specific and intrinsic viscosity of CA containing additives after irradiation by mercury-quartz lamp PRK – (2% of PAC)

CA – film, containing additives	Before irradiation		After irradiation		
	η_{spec}	H_{intr}	H_{spec}	H_{intr}	Conservation %
Initial	0,76	1,228	0,21	0,393	32,0
BA	0,76	1,228	0,25	0,462	37,0
XXXIV	0,78	1,254	0,54	0,923	73,6
XXXV	0,78	1,254	0,66	1,093	87,1
XXXI	0,78	1,254	0,64	1,072	85,4
XXXII	0,76	1,228	0,65	1,099	87,8
XXXVI	0,80	1,279	0,62	1,038	87,1
XXXIII	0,76	1,228	0,62	1,038	84,5

Figure 2.20 presents data, characterizing kinetics of specific viscosity conservation, and Figure 2.21 present data on kinetics of CH_3COOH accumulation in initial and stabilized CA – films at ultraviolet irradiation. As it is seen from these figures, intensive accumulation of acetic acid and decrease of specific viscosity is being observed in initial CA during the process of ultraviolet irradiation. But introduction of BA, XXXIV, XXXV additives inhibits these phototransformations, but to a different extent.

If the introduction of BA (curve 2, Figure 2.20 and 2.21) only slightly influences changing of specific viscosity and accumulation of acetic acid, then introduction of XXXIV and XXXV (curves 3 and 4 in Figure 2.20 and 2.21) facilitates considerable decreasing of acetic acid accumulation and increasing of indices of viscosity. So, after introduction of low-molecular BA initial specific viscosity is conserved by 16,7%, but addition of XXXIV and XXXII provides 50-80% of viscosity conservation. It should be noted that light-protective efficiency of XXXV (molecular mass = 1000) is much higher than that of XXXIV (molecular mass = 600).

The same picture is observed at CA – films irradiation by the lamp BUV – 30 (Figure 2.21). In this case there is kinetic dependence of acetic acid accumulation both in initial CA – film and in stabilized one. It is seen from Figure 2.21 that introduction of XXXV (curve 2) decreases acetic acid accumulation by 81,5% in comparison with initial CA – film.

A little another character of curves run is observed during irradiation by combined light of mercury-quartz and carbon-arc lamps. During irradiation of the initial CA, as in the previous case, linear dependence of acetic acid accumulation (Figure 2.22, curve 1) is observed. However, induction period is being observed after introduction of azomethines BA and XXXV (Figure 2.22, curves 2 and 3) at the initial phase of irradiation.

During further irradiation considerable increase of acetic acid accumulation takes place in CA-film containing BA and linear sector, beginning after 7 hours of irradiation is being observed. In the case of the presence of XXXV in the film induction period quickly increases. At the same time acetic acid accumulation with the increase of irradiation up to 10 hours is 8% of the same value of initial CA-film and 42% of the film, containing BA.

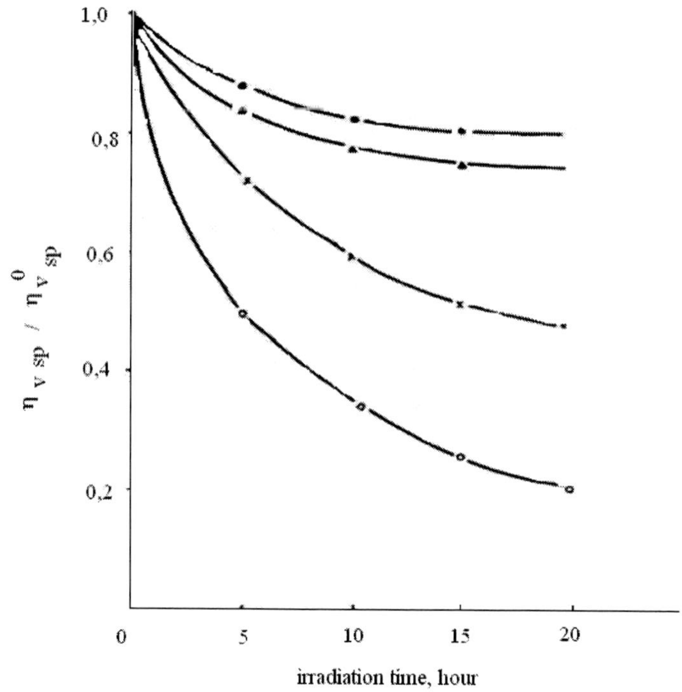

Figure 2.20. Change of specific viscosity of CA films in the process of irradiation by mercury-quartz lamp. 1 – CA-film without additive; 2 – CA-film, containing BA; 3 – CA-film, containing XXXIV; 4 – CA-film, containing XXXV.

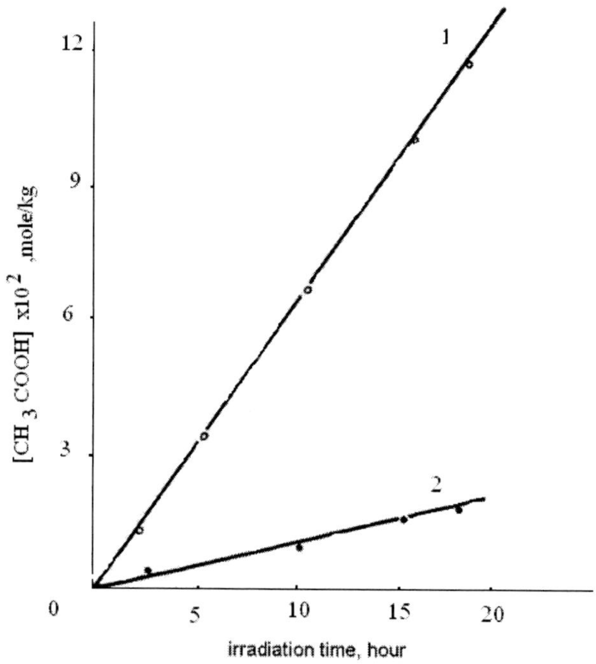

Figure 2.21. Change of the rate of acetic acid accumulation in the process of irradiation by carbon-arc lamp. 1 – CA-film without additive; 2 – CA-film + 2% XXXV.

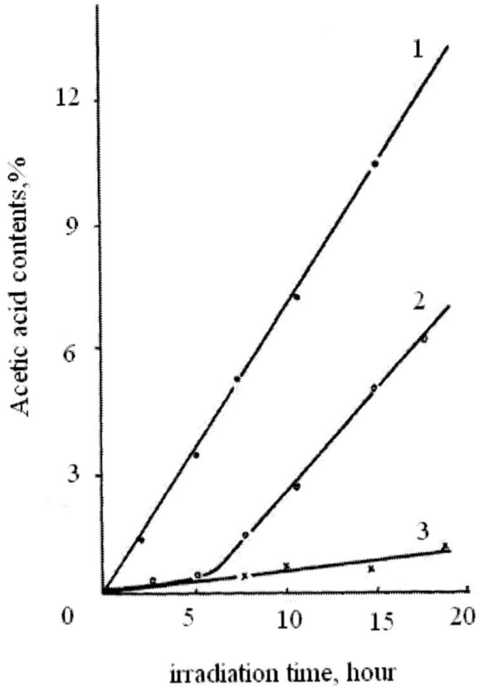

Figure 2.22. Change of the rate of acetic acid accumulation in the process of irradiation by carbon-arc and mercury-quartz lamps. 1 – CA-film without additive; 2 – CA-film + 2% BA; 3 – CA-film + 2% XXXV.

At natural insolation of CA-films kinetic dependences of acetic acid accumulation have similar character. Acetic acid of nonstabilized cellulose acetate accumulates more quickly than with BA and XXXV additives (Figure 2.23).

It should be noted that in all cases the largest inhibitive effect both according to viscosity values and acetic acid accumulation, is shown by XXXIV and XXXV, whereas low-molecular azomethine BA possesses slight light protective effect, which quickly decreases with the increase of irradiation time.

Destruction of ester groups was observed on the change of infrared spectrum in the ranges 1040 and 1230 cm^{-1}. Comparison of kinetic of the change of absorption band intensity at 1040 and 1230 cm^{-1} in infrared spectra of irradiated and nonirradiated by ultra-violet light CA-film in the air shows that as a result of ultra-violet light action complex destructive processes take place in CDA, visual display of which is the decrease of absorption bands intensity in the ranges 1040 and 1230 cm^{-1}, corresponding to stretching vibrations of ester bonds –C-O- (groups III) and change of carbonyl and aldehyde groups content. Curves run in Figure 2.24 correlates with data of Figure 2.23, which shows uniform nature of CA phototransformation at ultra-violet irradiation and dependences of stabilizing effect of azomethine compound on its molecular weight.

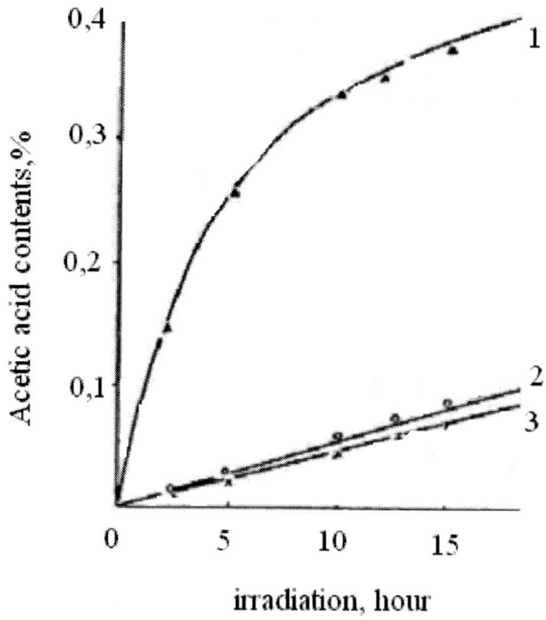

Figure 2.23. Change of the rate of acetic acid accumulation in the process of irradiation by sunlight. 1 – CA-film without additive, 2 – CA-film + 2% XXXV.

It is seen from the data in Figure 2.24 that intensity of absorption bands at 1040 and 1230 cm^{-1} decreases considerably during 10 hours irradiation. This indicated that at irradiation of CDA-films by mercury-quartz lamp splitting of ester bonds of acetyl groups takes place. But introduction of XXXIV and XXXV oligomer products, containing different number of conjugated bonds and also end amino-groups, into CDA considerably reduces number of splitting acetyl groups while low-molecular BA product, which does not contain amino-group, slightly influences this process. May be the last condition is described by the fact that shielding effect of PAC depends on the level of conjugation. That is why introduction of oligomer products XXXIV and XXXV with a large number of conjugated bonds shields the effect of ultra-violet light quantums on CDA ester groups, which is proved by the data on optical densities of the bands 1040 and 1230 cm^{-1}.

However, there is possibility that the presence of end amino-groups and also the last consumption of PAC at prolonged irradiation influence stabilizing properties of oligomers. Kinetics of optical density change of stabilized CA-films at absorption of additives (Figure 2.25) may help to judge about consumption of BA. From this figure it follows that low-molecular BA is consumed faster than oligomers XXXIV and XXXV and this also confirms more powerful and longer PAC action.

Data on carbonyl and aldehyde groups accumulation in initial and stabilized CA-films during the process of irradiation by mercury-quartz lamp are given in Figure 2.26. From this figure it is seen that carbonyl group accumulation is characterized by linear dependence. Intensive accumulation of carbonyl and aldehyde groups is observed in initial CA-film while slopes of strait lines of stabilized azomethine films (Figure 2.26, curve 2-6) are much smaller and this shows decrease of the rate of these groups accumulation. It is interesting to note that in this case BA also facilitates the decrease of carbonyl and aldehyde groups accumulation, though oligomer XXXVII displays the greatest activity in this respect.

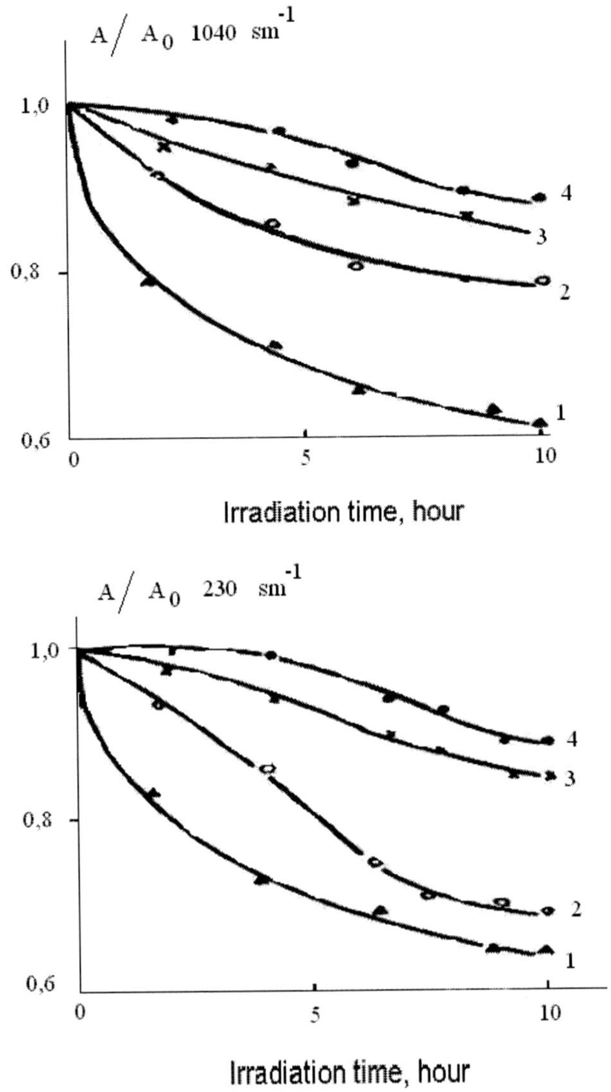

Figure 2.24. Kinetic of the change of absorption baunds intensity of CA-film in the process of irradiation by mercury-quartz lamp. 1 – CA-film without additive, 2 – CA-film with BA additive, 3 – CA-film with XXXIV additive, 4 – CA-film with XXXV additive.

Durability of polymer has been studied as the criterion of the effect of oligomer and polymer schiff's base on the mechanical properties of CA-film in the process of ultra-violet irradiation.

It is known that destruction of solids, specifically polymers, at any loading operation may be considered proceeding from general ideas on the nature of temperature-time dependence of strong solids. According to these ideas destruction is the kinetic process developing in the body under the load. The process of destruction means that the principle of disturbance summation should be kept. This principle may be expressed by the formula:

$$S\ dt/\tau[\sigma(t)] = 1 \qquad (10)$$

where τ^1 – time from the moment of application of load up to the break of a sample; $\tau[\sigma(t)]$ - dependence of durability on stress and temperature.

This dependence may be presented in the form:

$$\tau[\sigma(t)] = \tau 0 \cdot e^{[Uo - \gamma\, \sigma(t)]/RT} \quad (11)$$

where τ – time from the moment of application of load; U_0 – energy of activation of structure elements change in the sample under stress; γ – structure-sensitive coefficient.

CA-film with the additives of BA and schiff's bases XXXIV, XXXV, XXXVI and without additives was irradiated by mercury-quartz lamp at 50°C and irradiation intensity equal to 4,40 J/cm^2·min during 10 hours.

Quantitative value of coefficient γ, being defined by the tangent of the slope of dependence of durability logarithm 1gτ on the stress σ may be used as characteristic for evaluation of light resistance.

Growth of γ for irradiated samples in comparison with non-irradiated ones characterizes the degree of their light resistance decrease. Data on coefficient γ change in CA-films with and without additives are given in Table 18.

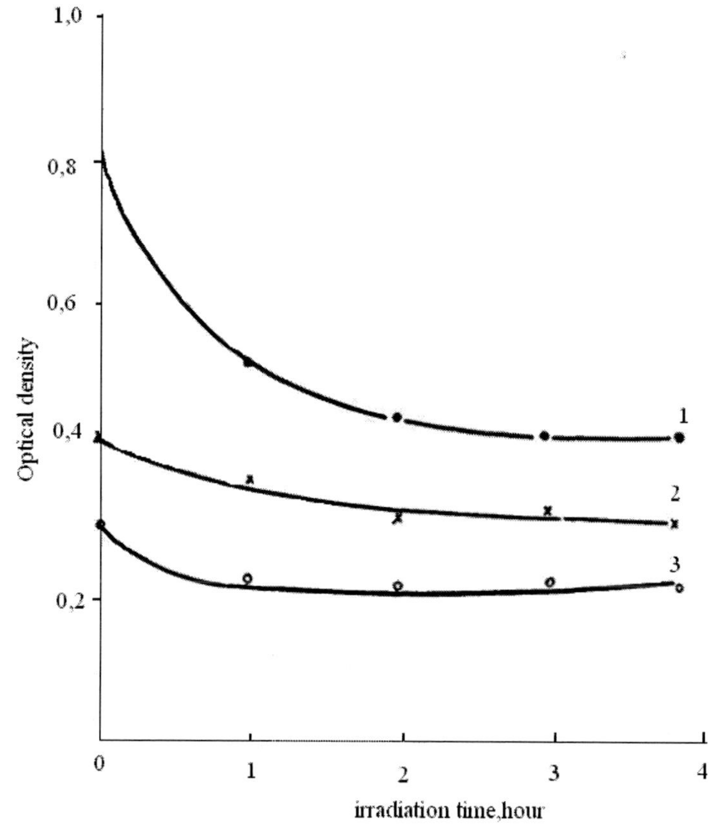

Figure 2.25. Kinetic of optical density change in CDA-film containing azomethines at corresponding Λmax: 1-330nm (BA), 2-272nm (XXXIV), 3-368nm (XXXV) in the process of irradiation by mercury-quartz lamp.

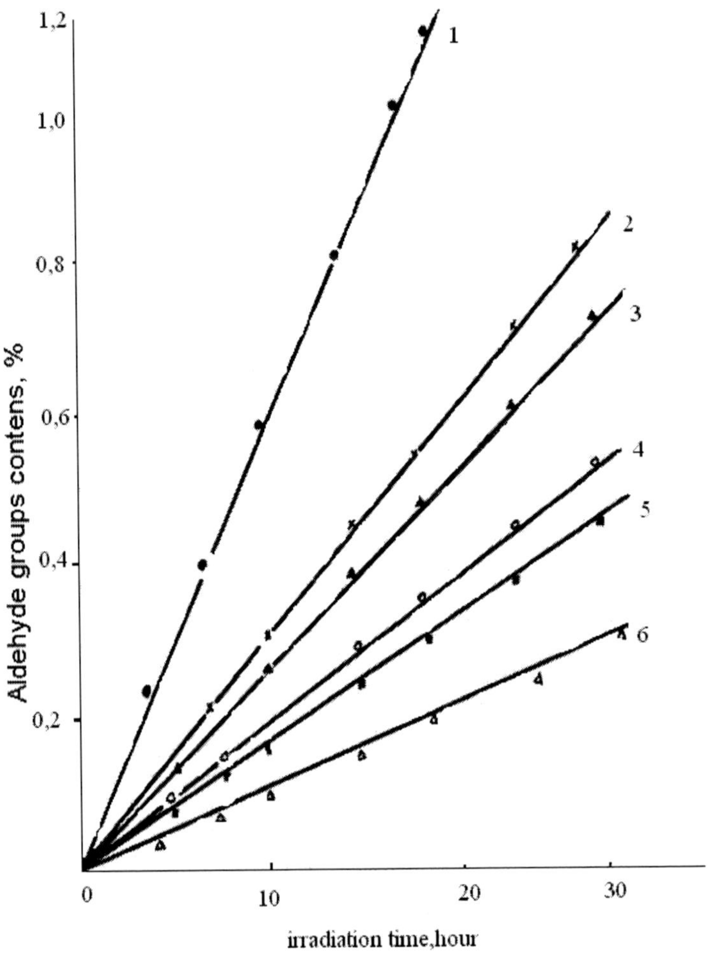

Figure 2.26. Kinetic of carbonyl groups change in CA-film containing 2% BA and PAC in the process of irradiation by mercury-quartz lamp: 1 – CA without additive, 2 – CA+BA, 3 – CA+XXXIV, 4 – CA+XXXV, 5 – CA+XXXVI, 6 – CA+XXXVII.

Table 18. Change of coefficient γ depending on the extent of conjugation and concentration of additives in CA at irradiation by mercury-quartz lamp during 10 hours

CA-film, containing	Molecular mass	Concentration of additives, %	Change of coefficient γ mm/kg	
			Before irradiation	After irradiation
Initial	-	-	2,68	5,66
		1	2,68	5,66
BA	181	3	2,68	4,50
		1	2,68	3,46
XXXIV	400	3	2,68	2,85
		1	2,68	3,07
XXXV	1000	3	2,68	2,68
XXXVI	1680	3	2,68	2,68

It is seen from Table 18 that coefficient γ has different values depending on the extent of PAC conjugation, being introduced into cellulose acetate. So, initial CA and CA with BA additive after irradiation have the largest value of γ. Coefficients γ in CA with XXXV and XXXVI additives after irradiation do not change considerably in comparison with the value of initial CA. These results show that light resistance of CA stabilized samples essentially depends on the extent of PAC conjugation.

Data of Figure 2.27 according to durability both initial and stabilized samples of CA-films are the visual confirmation to information mentioned above. It is seen from Figure 2.27 that experimental data on durability of both initial and stabilized non-irradiated samples are on the same line of dependence.

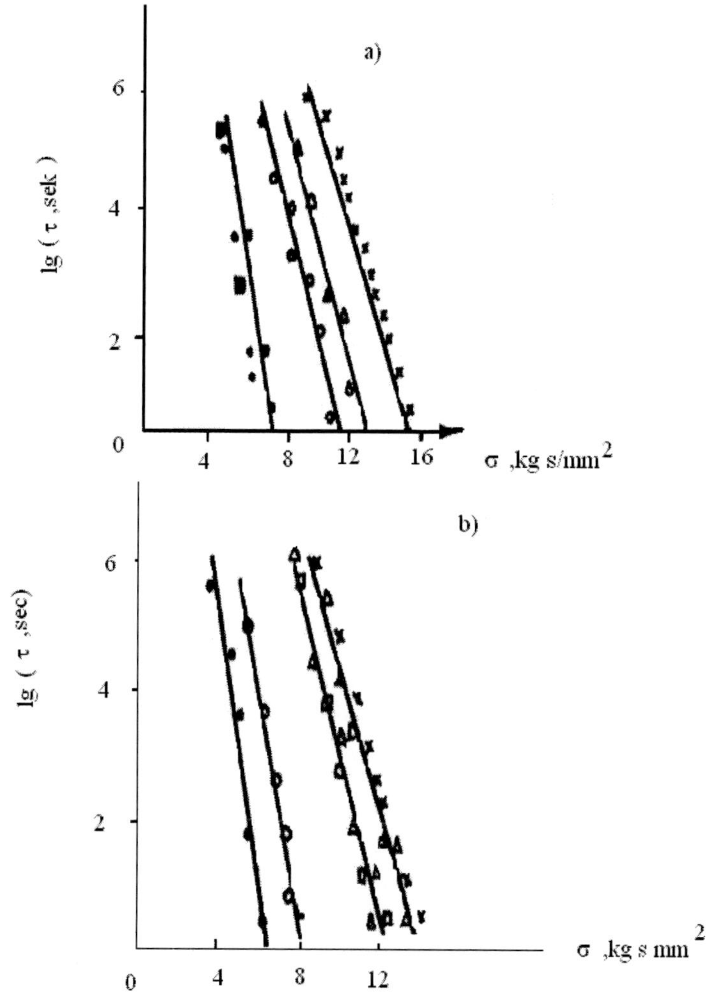

Figure 2.27. Effect of preliminary ultra-violet irradiation on CA durability. Additives in %: a – 1, b – 3. o – non-irradiated CA, x – CA+BA irradiated, A –CA+XXXIV irradiated, A – CA+XXXV irradiated, Q – CA+XXXVI irradiated, XXXVII – irradiated CA.

This fact means that introduction of light-stabilizer 1-3% of polymer mass into CA-film does not practically effect on its initial strength till irradiation. Low-molecular additive BA in concentration 3% of polymer mass does not influence CA durability after irradiation as well, whereas introduction of XXXV and XXXVI almost completely conserves durability of irradiated polymer.

Experimental data show that efficiency of light stabilization at photoageing depends on the extent of conjugation of introduced CA-product-PAC [180].

On the basis of above mentioned it may be assumed that effect of light stabilization is probably linked with the increase of electronic excitation transfer from polymer to PAC. And it may be expected that the more complex is PAC molecule the more vibrating degrees of freedom it has and the more is the observed effect of excitation suppression [180].

2.7. Light Stabilization of CDA By Nitrogen and Sulphur Containing Aromatic Compounds

It is known that aromatic compounds containing heteroatoms of nitrogen may have light stabilizing effect at polymers irradiation. Some derivates of carbazole, according to literature and patent data, may be CC and antioxidants during oxidative destruction of polymers.

So, it seemed interesting to study the influence of substitutes in sulphuryl amides concerning light stabilizing effect intensity. For this purpose we modified cellulose diacetate (CDA) with the help of sulphuryl amide derivatives of carbazole and indan.

Among sulphuryl amide derivatives of indan the first to be synthesized was 5-indan sulphonyl chloride - a result of indan and chlorosulphonic acid interaction. Then amines of benzene series were acetated by previously produced 5-indan sulphonyl chloride. Some of the appearing sulphuryl amides came through alkylation by dimethyl sylphate in sulphuryl amide group. Carbazole derivatives were obtained by acetating of aniline with special carbazole sulphanilchlorides according to the method [181].

These synthesized sulphuryl amide were used for cellulose diacetate (CDA) modification. The obtained samples were formed as films and radiated by mercury vapour lamp in the air during the period of 24 hours at T 35^0C. Change of viscosity of cellulose diacetate solution before and after irradiation was considered as the index of stabilizing effect of additions – modifiers[182].

Tables 19 and 20 display the experimental results of sulphuryl amide derivatives of carbazole and indan stabilizing effect.

Obtained data show high light-protective activity of studied compounds, which, perhaps, may be either compared or exceed the activity of Tinuvine II.

Taking into account the fact that carbazolsulfonamide XXI-XXVIII (Table 19) absorb in the same field as in CDA, then it may be supposed that stabilizing effect is caused by “shielding” action.

In order to elicit the light stabilizing mechanism of the examined additions, modified CDA- films were dissolved in acetonitrile before and after being radiated. The obtained solutions were spectrographed in Ultra-violet (Figure 2.28-2.30).

Table 19. Viscosity maintenance in CDA- films containing carbazole sulphuryl amide additions in ultra-violet radiation

compound No.	Formula	Structure	Viscosity in %		
			Concentration of a stabilizator in % to polymer mass		
			0,5	2,0	5,0
XXI	$C_{13}H_{12}SO_2N_2$	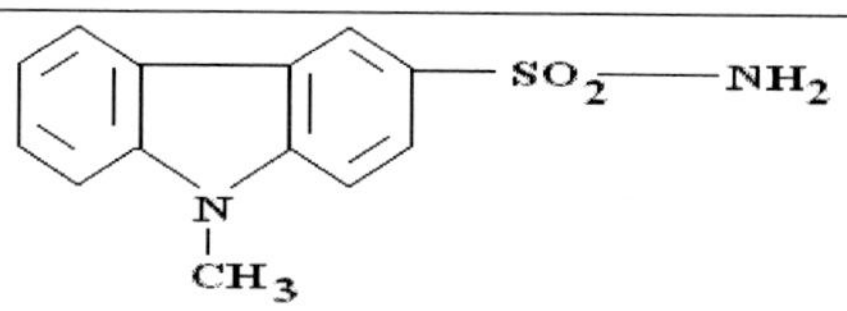	-	63,0	-
XXII	$C_{14}H_{12}SO_3N_2$	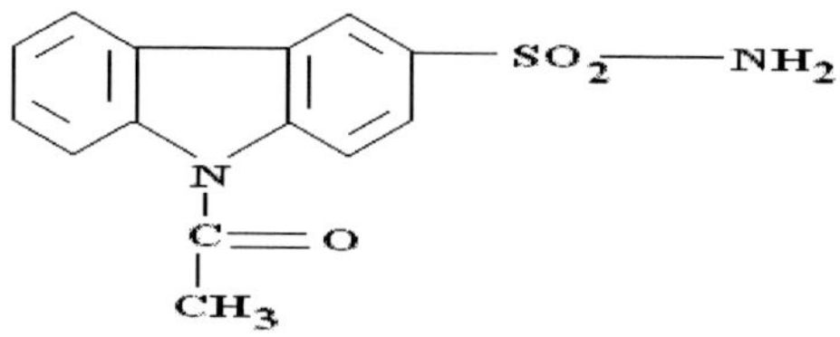	-	74,0	-
XXIII	$C_{20}H_{18}SO_2N_2$	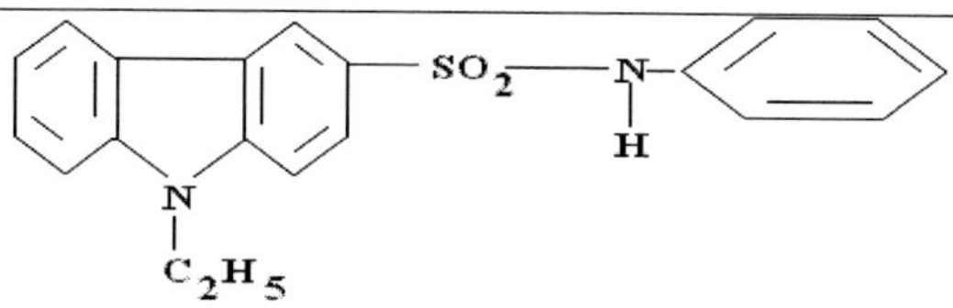	51,0	55,3	64,5
XXIV	$C_{18}H_{13}SO_2N_2Cl$	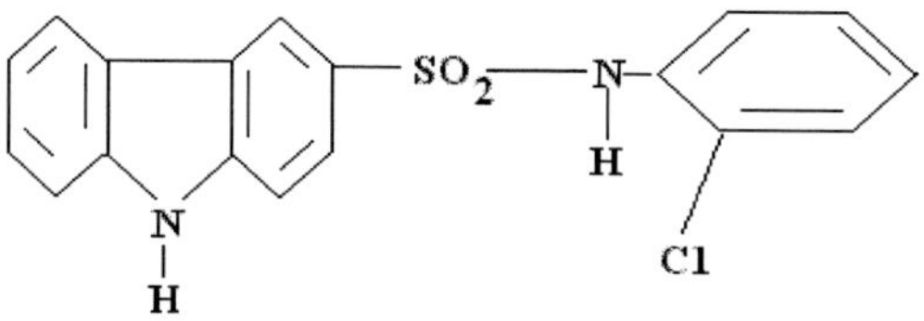	53,5	-	82,9
XXV	$C_{18}H_{13}SO_4N_3$	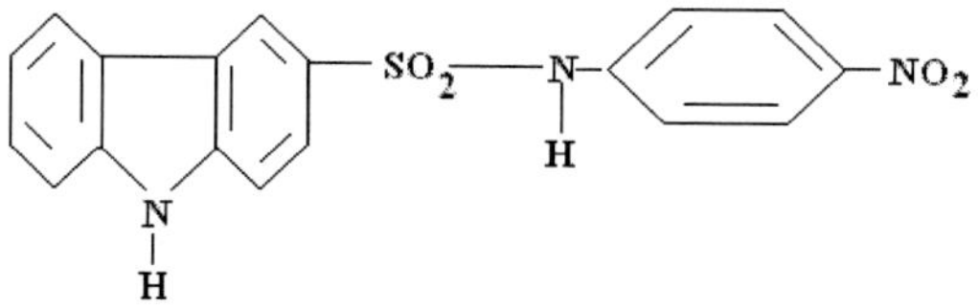	56,9	68,8	83,1

Table 19. Continued

compound No.	Formula	Structure	Viscosity in %		
			Concentration of a stabilizator in % to polymer mass		
			0,5	2,0	5,0
XXVI	$C_{18}H_{14}SO_2N_2$	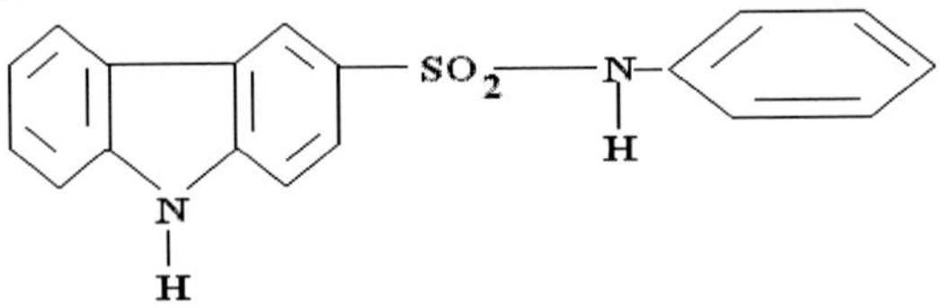	20,5	60,0	72,4
XXVII	$C_{21}H_{20}SO_2N_2$	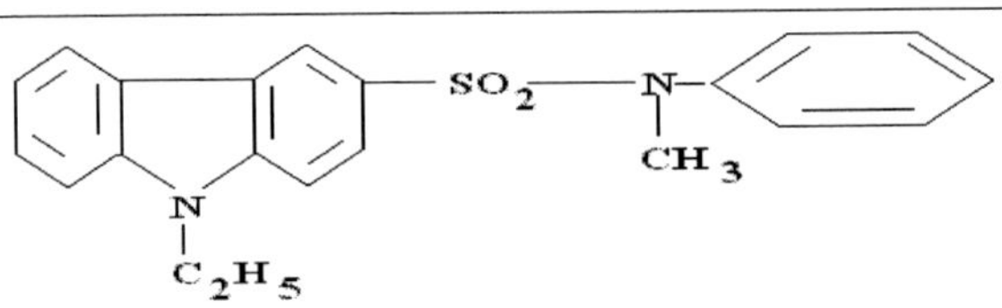	58,4	55,8	62,1
XXVIII	$C_{16}H_{18}SO_2N_2$	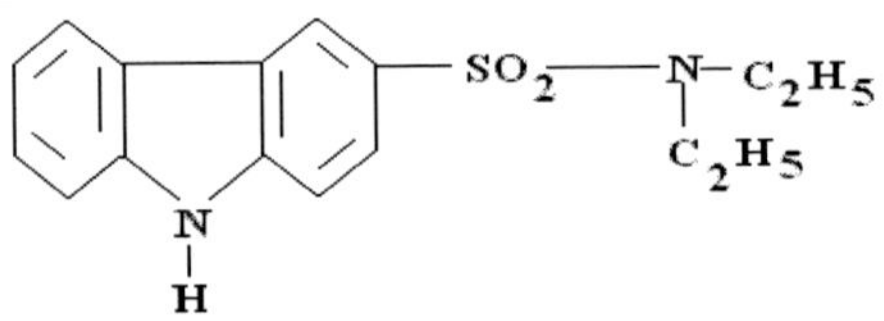	21,9	20,9	47,7
XIX	$C_{24}H_{18}SO_2N_2$	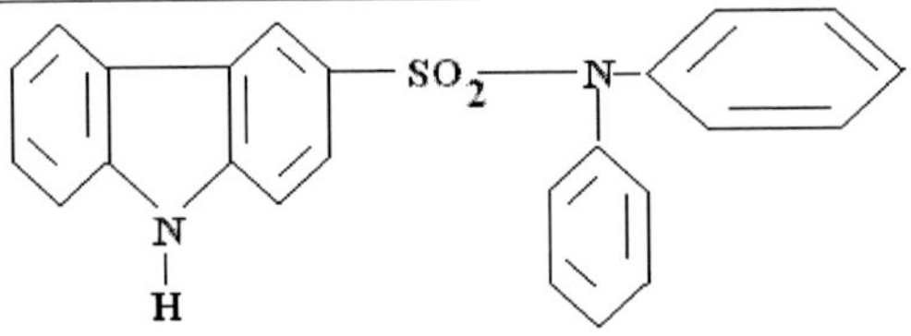	-	42,8	-
XXX	$C_{19}H_{16}SO_2N_2$	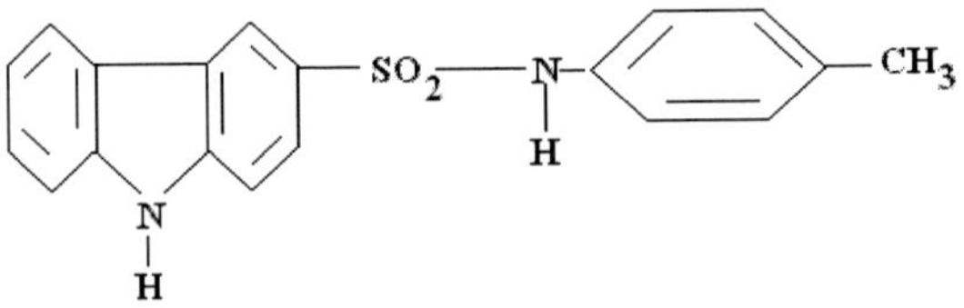	46,2	55,4	59,4
Tinuvine-II	-	-	-	34,4	-

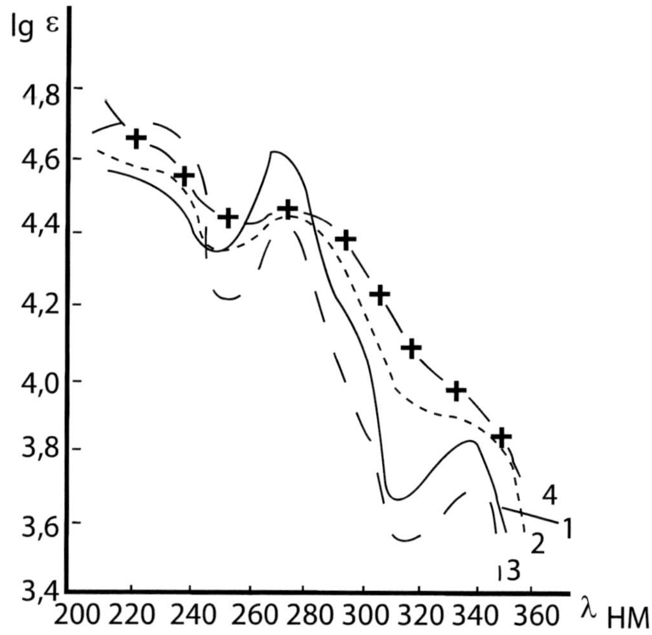

Figure 2.28. U-v spectrum (acetonitrile) 1-CA-film, containing 5 % of compound (II) before irradiation; 2 – the same sample after irradiation; 3 – CA-film, containing 2% of compound (II) before, 4 – and after irradiation.

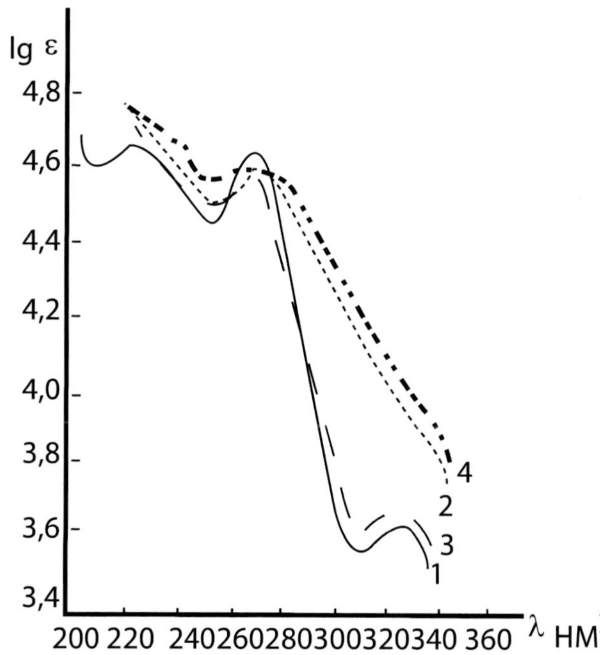

Figure 2.29. U-v spectrum (acetonitrile) 1- CA-film, containing 5 % of compound (III) before irradiation; 2 – the same sample after irradiation; 3 – CA-film, containing 2% of compound (III) before, 4 – and after irradiation.

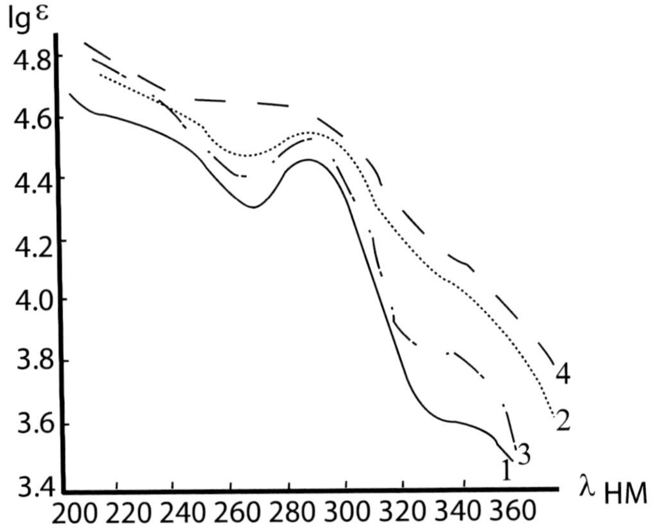

Figure 2.30. U-v spectrum (acetonitrile) 1- CA-film, containing 5 % of compound (V) before irradiation; 2 – the same sample after irradiation; 3 – CA-film, containing 2% of compound (V) before, 4 – and after irradiation.

Table 20. Kinetic parameters of polymerization in the presence of carbazolsulphonamides

№	$W^{st}_{int} \cdot 10^4$, mole/l·s	F_{st}	$W^{gel}_{int} \cdot 10^4$, mole/l·s	F_{gel}
XXIII	5,0	0,64	41,3	0,83
XXVII	3,8	1,28	23,0	2,32
XXX	4,7	0,77	34,7	1,32
XXV	6,0	0,28	42,2	0,79
XXVIII	3,5	1,49	32,8	1,36
without inhibitor	6,9		62,0	

Measuring U-v – spectrum displayed, that modifier concentration of 2% to the mass of a fiber change spectrum considerably. It is necessary to note, that after irradiation the spectrum looks differently: absorption bands change to reduction. Possibly, these changes are caused by absorption of CDA decomposition products CDA, which, in its turn, decelerates further polymer decomposition.

As it is established in [183], aromatic sulphuryl amides can exist in solid state in the form of open or cyclic dimers owing to intermolecular hydrogen bonds.

There is an opinion, that sulphuryl amide grouping can have anti-configuration, that is, either O-S-N-H atoms are lying in the same plane or oxygen and nitrogen atoms oppose each other [184].Thus, as carbazole and indan derivatives include sulphuryl amide grouping, it is possible that light stabilizing effect of additions is due to hydrogen bonds, as it happens in o-hydroxybenzophenone [185]. The light stabilizing effect of the latter is determined by rapid and convertible phototransition of a proton participating in intermolecular hydrogen bond formation.

At this same time there are data that arylsulfanamides may inhibit radical-chain process of destruction. Kinetics of inhibited radical polymerization of methyl methacrylate, initiated by dinitrile of azo – bis – isobutyric acid, was studied to evaluate inhibiting activity of carbazolsulphonamides.

Inhibiting ability of carbazolsulphonamides was characterized by change of the rate of methyl metacrylate polymerization in the presence of suggested compounds in the mode of the stationary flow of the process (W_{int}^{st}) and "gel-effect" (W_{int}^{gel}) and also by the factor of inhibiting (F), the value of which is proportional to the constant of the rate of inhibition.

As it is seen from kinetic parameters, given Table 20, all carbazolsulphonamides on way or another decrease the rate of polymerization, however, induction period here is absent, hence, these compounds are weak inhibitors. May be additional shielding effect after addition of carbazolsulphonamides in connected with their participation in inhibition of radical processes of CDA photodestruction.

Tinuvine –II, which, as it is known, is not an inhibitor of radical processes, appeared to be less active stabilizer of photooxidative destruction.

To compare the effect of "carbazole" component of sulphonamides on light stabilizing activity there has been carried out their synthesis on the basis of coke-chemical indan (Table 21).

Table 21. Viscosity maintenance in CDA- films containing indan sulphuryl amide additions in ultra-violet radiation

No. of compounds	Formula	Structure	T, ^{0}C	Viscosity, %
X	$C_{15}H_{15}NO_2S$	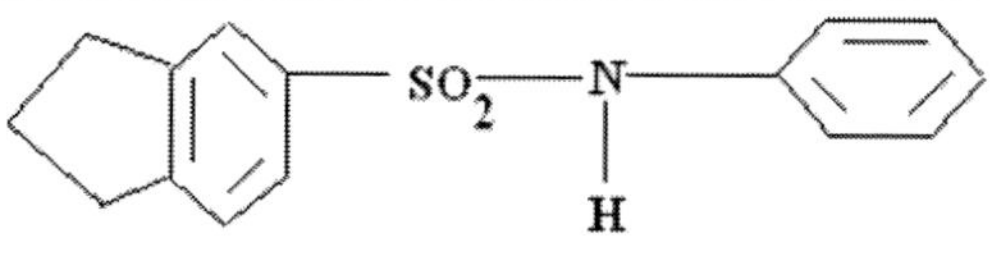	135-136	36,6
XI	$C_{15}H_{15}NO_3S$	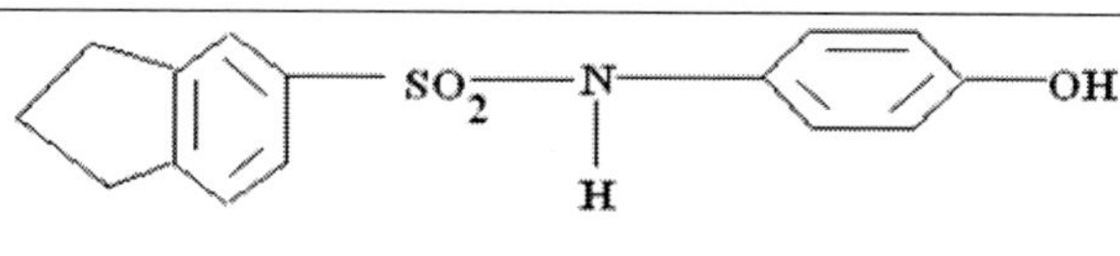	158-159	42,2
XII	$C_{16}H_{17}NO_2S$	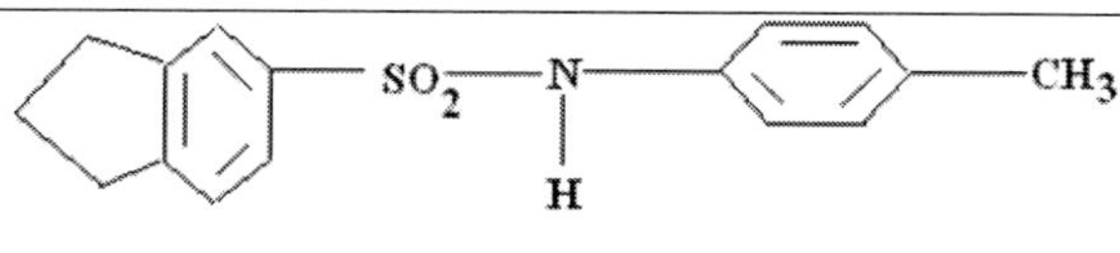	144-146	32,2
XIII	C15H14N2O 4S	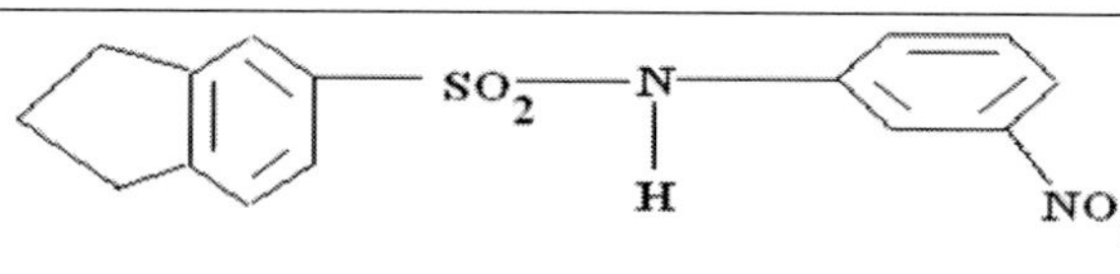	185-186	47,7

Table 21. (Continued)

No. of compounds	Formula	Structure	T, ^0C	Viscosity, %
XIV	$C_{15}H_{16}N_2O_2S$	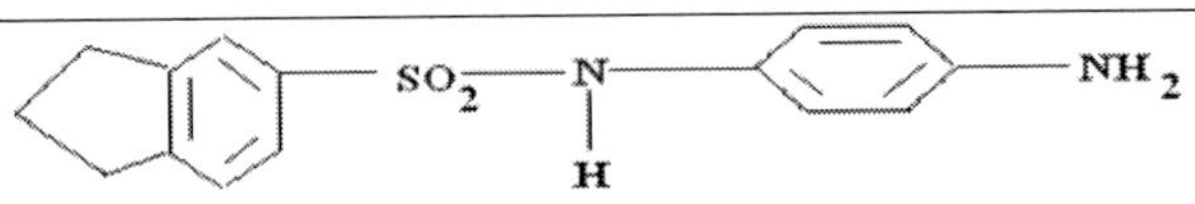	230 destr.	24,0
XV	$C_{24}H_{24}N_2O_4S_2$	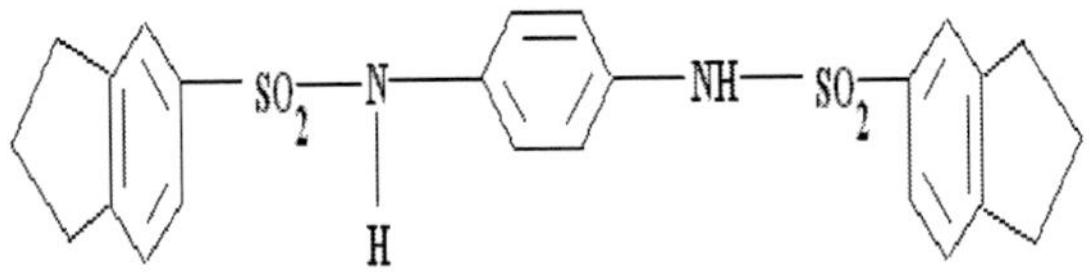	280 destr.	29,1
XVI	$C_{16}H_{17}NO_2S$	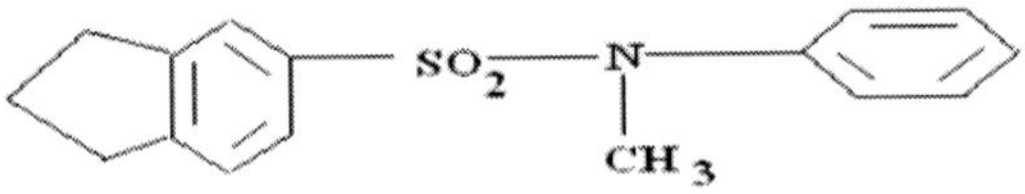	95-96	40,7
XVII	$C_{17}H_{19}NO_2S$	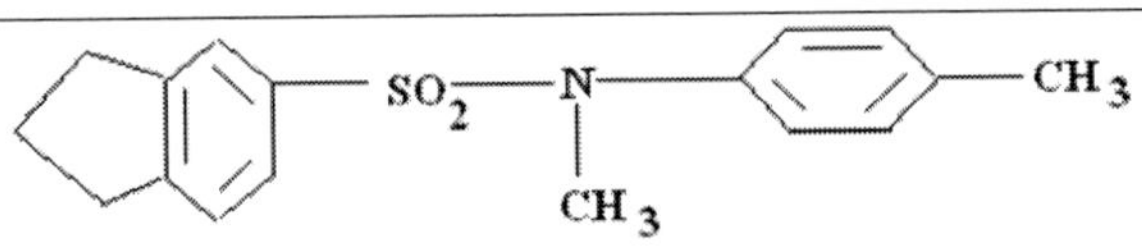	102-103	33,3
XVIII	$C_{16}H_{17}NO_3S$	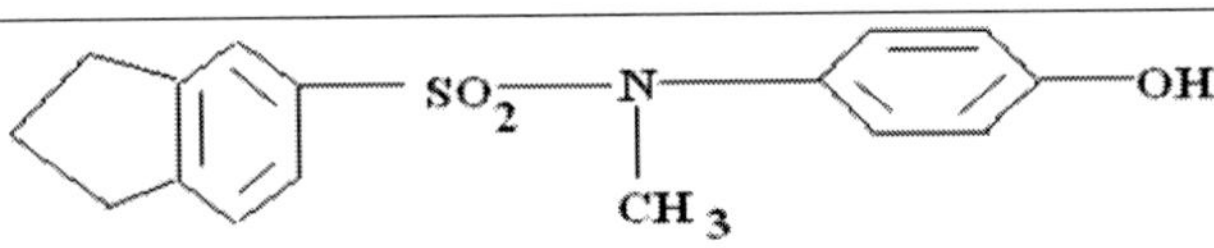	120-121	37,9
XIX	$C_{16}H_{16}N_2O_4S$	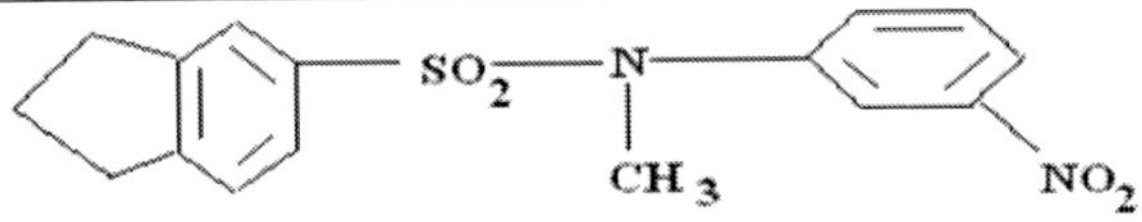	196-197	47,1
XX	$C_{16}H_{18}N_2O_2S$	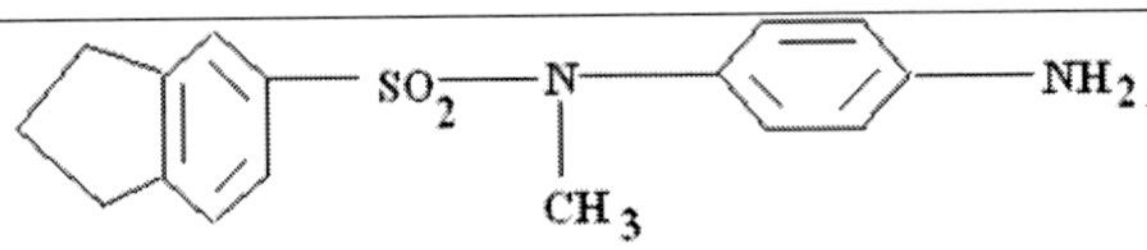	196-197	62,2

Studying ultra-violet spectrums of indan derivatives let us make a supposition about their "shield" mechanism, which causes light stabilizing effect, as indan sulphuryl amide additions (as carbazolsulphonamides) absorb radiation in the same wave length interval as cellulose diacetate.

As it is seen from Table 21 all indan - sulphonamides facilitate CDA photo stabilization. It may be noted that, with some exception, introduction of different substituents into indan - sulphonamides greatly influences their stabilizing activity.

In the case of amino- indan - sulphonamides XIV and XX there is observed sharp difference in light protection effect (24,6 and 62,6% respectively), though these compounds are distinguished only by presence of methyl group, linked with nitrogen of sulphamid bridge. This difference is well described by ultra-violet spectra of stabilized CDA-films non-irradiated and irradiated during 24 hours (Figure 2.31 -2.32.).

Indan - sulphonamide XX has maximum absorption at 256 nm, whereas compound XIV – at 232 nm, which demonstrates stronger effect of conjugation in XX. Besides, in irradiated CDA-film the ratio of absorption intensity to the initial value at corresponding λ_{max} in compound XIV is 0,68, whereas in XX this ratio is 0,86. It shows that stabilizer XIV during irradiation is consumed 1,3 times faster than stabilizer XX. May be methylation of nitrogen in sulphamid group stabilizes the structure, which, in its turn, increases intensity and duration of light stabilizing effect.

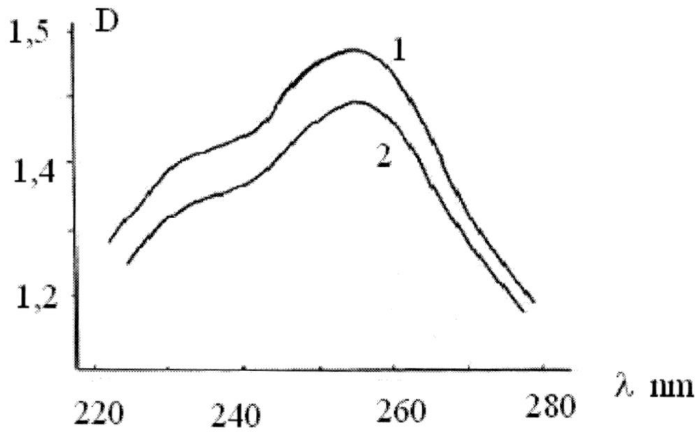

Figure 2.31. U-v spectrum of absorption in CDA - films, containing 2% of XX indan sulphuryl amide to polymer mass: 1- CDA - film before irradiation; 2-after 24 hours u-v – irradiation.

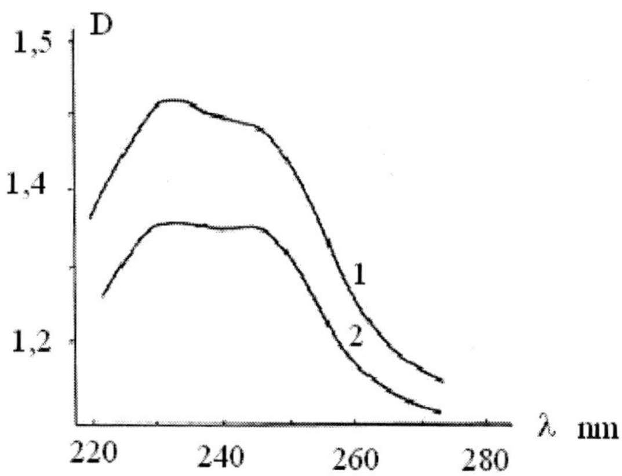

Figure 2.32. U-v spectrum of absorption in CDA - films, containing 2% of XIV indan sulphuryl amide to polymer mass: 1- CDA - film before irradiation; 2-after 24 hours u-v – irradiation.

It is very important to note that indan - sulphonamides absorb light in the same field as CDA. The last fact allows to make a conclusion about "shielding" mechanism of the action of this set of compounds.

Comparison of data on light stabilizing activity of indan - sulphonamides with results, obtained for sulphonamides on the basis of carbazole, shows that substitution of indan fragment for carbazole cycle leads to the increase of conjugation effect and, as a result, to the increase of light protective effect by 15-26%(Table 22).

Light-protective effect changes noticeably depending on substituent both in aromatic cycle and in amine component. So, carbazol-sulphonamides, obtained on the basis of aliphatic amines, display much lower activity than arylamine derivates. Quick increase of light stabilizing activity is observed during introduction of electronoacceptor substituents (R_a) into phenylene rings and, on the contrary, electrono-donor substituents slightly reduce protective action of stabilizer.

Table 22. The comparative characteristics of stabilizing effect of sulphuryl amide derivatives of carbazole and indan

Compound	Structural formula	Viscosity, %
indan derivatives (X)	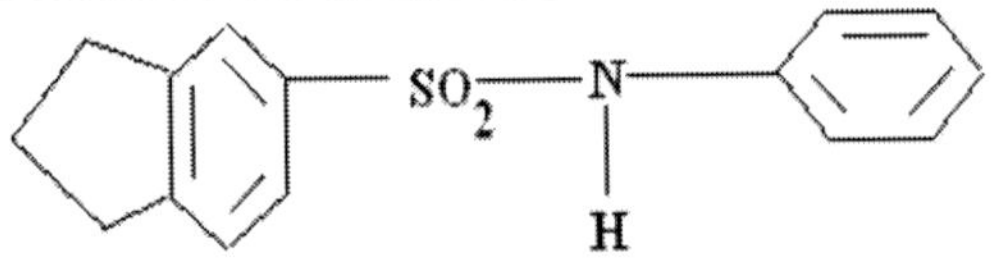	36,6
carbazole derivatives (XXVI)	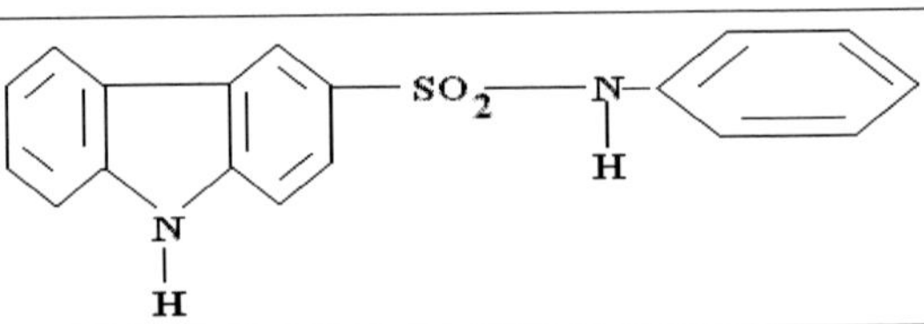	72,4
indan derivatives (XII)	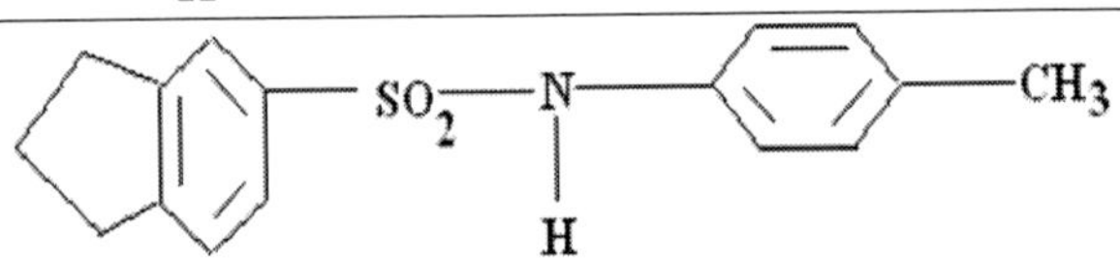	32,2
carbazole derivatives (XXX)	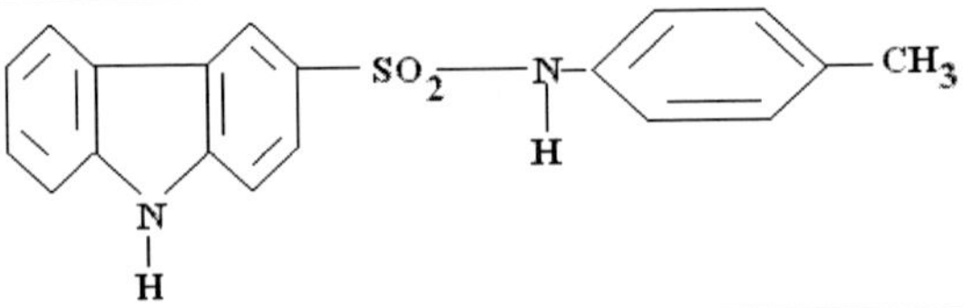	83,1
indan derivatives (XVI)	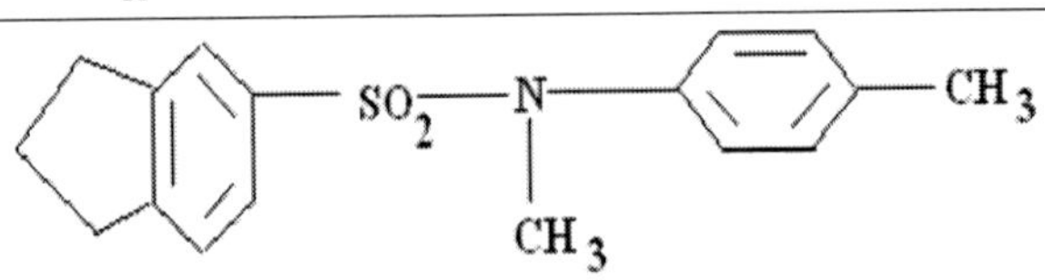	40,7
carbazole derivatives (XXVII)	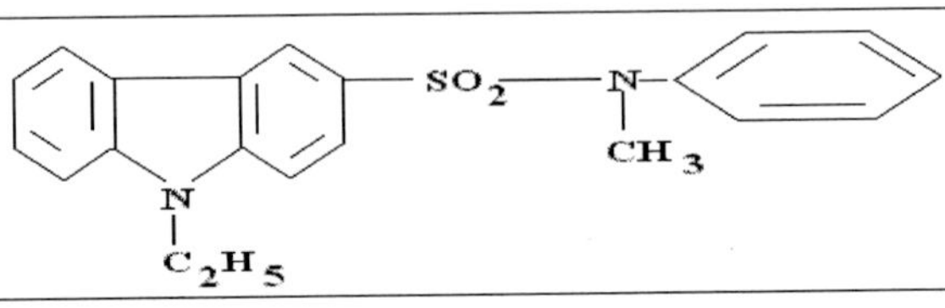	62,1

Substitution of hydrogen for alkyl residue in heteroatom of carbazole cycle causes high level photoprotective effect even at 0,5% stabilizer cocentration in polymer, the more so, this effect doesn't change considerably with futher increasing of concentration (Table 20 XXVI and XXIII compounds).

Inhibiting influence undergoes notable change depending on substituent in both carbazole cycle, and in amine component. Thus, carbazole sulphuryl amides derived from aliphatic amines (XXVIII, XXIII, XXII) show lower activity, than those of acrylamin derivatives (XXVIII, XXVII). Substituent intrusion into phenylen rings results in sharp increase of light stabilizing activity in carbazole derivatives, as well as in those of indan (XXVI and XXX compounds or XVI and XVIII compounds). Thus, intruding electron accepting substituents (-NO_2) couses greater stabilization of CDA, than penetration of electron donation particles (-CH_3, -OH, -NH_2). Possibly, it is connected with ability of electron accepting particles to draw forward electronic density of benzene ring, as a result, carbon atoms get positive charge and hydrogen connected with them becomes mobile.

Comparing of stabilizing activity among electron donation substituents allows us to notice, that the stronger electron donating properties are exposed, the greater is light stabilizing effect (compound XX – substituent –NH_2 (62,2%); compound XVIII – substituent –OH (37,9%), compound XVII- substituent –CH_3 (33,3%).

It is important to notice, that sulphuryl amides derivatives of both indan and carbazole have a number of valuable properties of practical application: they are resistant to thermical shock (T decomp =120-300^{0}C); non-toxic (LD_{50}=1000 mg/kg); well consistent with CDA in common solvents; acquire small volatility.

Screening of formerly unknown sulphur containing compounds on the basis of benzo/B/-thiophene for their light stabilizing activity was carried out with the purpose of stabilizers assortment expansion. Corresponding data of specific viscosity conservation of 0,5% acetone solutions of CDA with different content of benzo/B/-thiophene are given in Table 23.

Table 23. Ultra-violet irradiation effect on the conservation of specific viscosity of 0,5 CDA solutions

Stabilizer	Stabilizer concentration, % of polymer mass				
	1	2	3	5	10
I	-	37,1	-	-	-
II	-	63,3	-	64,0	61,1
III	44,0	55,6	86,2	-	-
IV	39,2	-	66,9	-	-
IV	36,1	-	58,6	-	-
V,VI	38,0	42,1	-	84,2	-
VII	-	56,8	-	-	77,8
VIII	24,3	47,5	76,7	82,4	-
IX	82,7	87,9	-	72,8	-
Tinuvine-II	-	34,4	-	-	-

Screening showed that derivatives of benzo/**B**/-thiophene effectively prevent photooxidative destruction of polymer, exceeding well-known industrial stabilizer. It has been found that increase of CC concentration does not cause considerable rise of light stabilizing effect. This shows "shielding" effect of sulphur containing compounds, though their action as oxidants cannot be excluded.

2.8. STABILIZING BY MEANS OF CHEMICAL MODIFICATION OF CDA

Earlier Kargin V. made a hypothesis that on interesting method of stabilization is chemical modification of polymer by addition of stabilizers according to active functional groups of high-molecular compound. And though, on practical part, this method is less promising, at is connected with the change of technological conditions of polymer production, in some cases it is worth of paying attention to, since stabilizer washing-out during wet treatments, its volatility in vacuum or at high temperature action are eliminated.

For this purpose CDA modification by its condensation with (diphenyl-p-trilisocynate) urea (DTIU) was carried out according to the reaction:

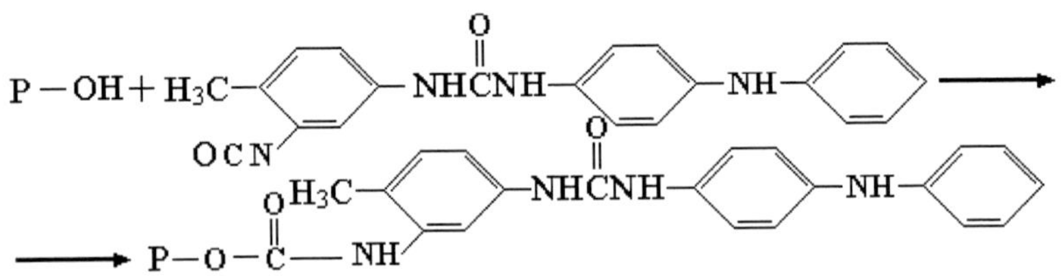

Before hand testing of light stabilizing activity of DTIU while introducing it into a polymer as mechanical additive was carried out. Data of Table 24 show that DTIU possesses photostabilizing activity that gives the reason to consider the fact that its condensation with CDA will improve light fastness of the latter.

Table 24. Effect of cellulose diacetate films ultra-violet irradiation on conservation of viscosity characteristics. Distance from the irradiation source is 30cm, time of exposure is 24 hours

DTIU content (% of polymer mass)	Specific viscosity of polymer η_{spec}		Degree of viscosity conservation (%)
	before irradiation	after irradiation	
0	2,00	0,46	29,0
2	2,00	1,46	73,0
4	2,00	1,60	80,0

Molecular weights of initial CDA and its modified analog were calculated according to viscosimetric data with the use of formula $[\eta]=KM^{\alpha}$, where $K = 0{,}19 \cdot 10^4$ and $\alpha = 1.03^4$. These molecular weights before irradiation were $M_{in} = 62660$ and $M_{mod} = 69020$ respectively, and after irradiation they were: $M_{in} = 23390$ and $M_{mod} = 61940$. The number of breaks after irradiation, calculated on the basis of these data, of the initial CDA was 2,68 and of modified analog it was 1,11.

Investigation of mechanical properties of modified and initial CDA before irradiation, as it is seen from Table 25, shows that they are similar in properties. However, irradiation exerts different effect on CDA and its modified analogs.

It follows from Table 25 that modification of CDA considerably increases resistance of this type of compounds to severe ultra-violet irradiation. Since light waves range 300-400nm is the most dangerous for CDA then it may be assumed that light stabilizing action of this compound is based on the absorption mechanism of light absorption. The fact of slight increase of DTIU stabilizing activity at considerable increase of its content in polymer attracts attention and this indirectly indicates inhibitive activity of DTIU during oxidative destruction of polymer and is proved by studying its thermal oxidation.

Table 25. Effect of ultra-violet irradiation on mechanical properties of the films of modified and initial CDA

DTIU content (% of polymer mass)	before irradiation		after irradiation		Degree of conservation (%)	
	mm2	%	mm2	%		
0	637,4	3,62	310,4	0,45	48,7	12,8
0,2	654,0	5,48	379,5	2,42	57,8	44,2
1,0	601,5	4,06		2,28	88,7	56,2
3,5	622,3	3,86	75,0	24	92,5	58,2

Investigation of the kinetic of mass loss at thermooxidative destruction of modified CDA showed that modification increases its stability to the heating at elevated temperatures (Figure 2.33). Loss of mass of modified CDA in much less in comparison with initial values during prolonged heating of polymer samples in the air (150°-200°C). These conclusions are proved by the results of complex thermogravimetric analysis (TGA).

From the curves of TGA it follows that temperature of decomposition of modified CDA is 20°C higher that of non-modified one. Presence of two exthermical peaks on the DTA curves of initial CDA and absence of such peaks of modified polymer at the temperatures 240°C and 290°C are seen here.

May be this difference is connected with the fact that the number of OH-groups, being subjected to dehydration during heating of initial CDA are mush larger than of modified one. Effective activation energy, calculated according to the method of Freeman-Karrol, of modified CDA was higher by 504 J/mole in comparison with initial polymer.

On the whole we may make a conclusion that expected effect of photo- and thermo-stability appeared to be slight.

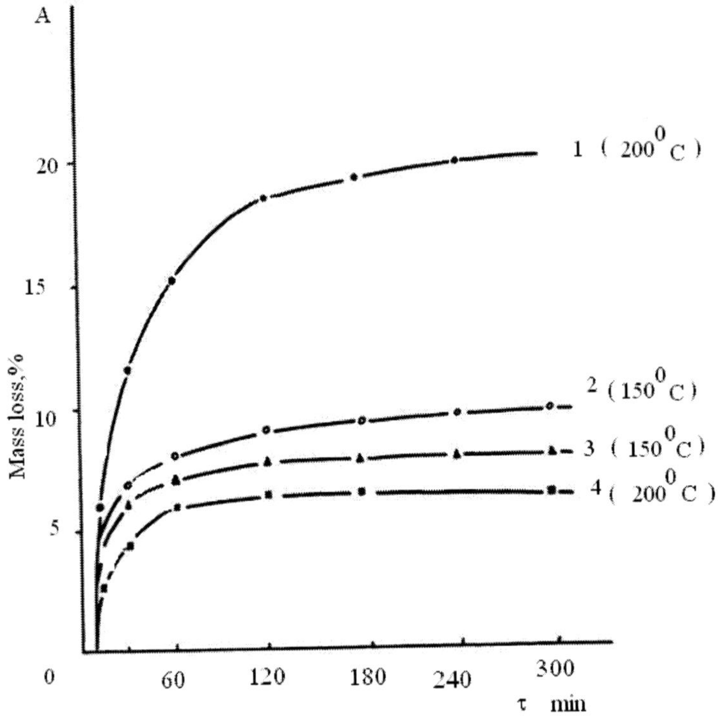

Figure 2.33. Curves of loss of mass of modified and initial cellulose diacetate under isothermal conditions. A – loss of mass (%), τ – time (min); 1,2 – initial cellulose diacetate at temperatures 200 and 150°C respectively; 3,4 – modified cellulose diacetate at temperatures 200 and 150°C respectively.

2.9. Thermo- and Photooxidative Destruction of Dyed Polyvinyl - Alcohol Fibres

Presence of considerable number of hydroxyl groups in polyvinyl alcohol (PVA) allows, with known degree of approximation, to consider this polymer as the model of cellulose and its derivatives.

Method of polymer materials protection by addition of low-molecular additives was used at modification of PVA fibres which in the future well allow to proceed to investigation of dyed materials on the basis of cellulose and its derivates with more confidence. PVA dyed by colours LIX, LXI, LXIII and their deactivated analogs LX, LXII, LXV were used in our work. DTA curves of PVA samples, dyed by colours LIX and LXI , and solid solutions LX and LXI in PVA in comparison with undyed analogs are shown in Figure 2.34. It should be noted here that dyed PVA-films do not have deep endoeffect at 120°C corresponding to the loss of sorption moisture, unlike initial film. Endoeffect at 220°C (initial PVA), which is not accompanied by the loss of mass and corresponding to meeting of polymer crystalline regions, is shifted in dyed samples into region of 235-239°C. Endoeffect at 280°C (curve 1 in Figure 2.34), characterizing the beginning of deep dehydration, is being observed in the case of solid solution LXII and LX in PVA already at 309-310°C and in the case of colour LIX covalently linked with PVA it is observed at 358°C, and of dyed LXI – at 341-349°C. Destruction of DTA curves is much more visible in the range of 400-500°C, where complex processes of oxidation and decomposition of PVA take place. DTG curves indicate that maximum rate of mass loss pf initial PVA is observed at 267°C. This index of dyed samples shifts in the direction of higher temperatures, and maximum of loss of mass of covalently linked dye LIX is at 375°C which is 100° higher than that of initial PVA. Data on the thermal stability of dyed PVA in comparison with undyed ones, obtained on the basis of TG curves are presented in Figure 2.35. These curves show that thermal stability of PVA after addition of dye increases as intensive destruction of dyed samples begins at higher temperatures and depth of destruction at one and the same temperature decreases. Dyes covalently linked with PVA, especially LIX dye, display the highest thermostabilizing activity.

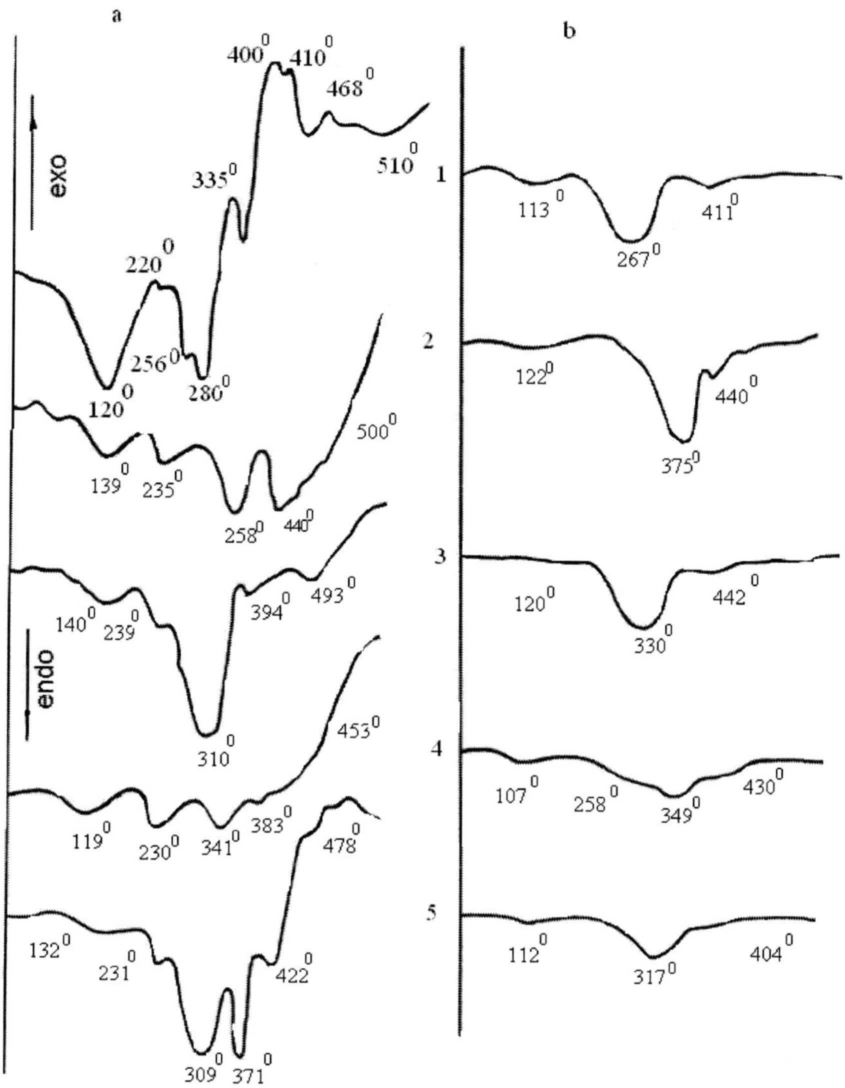

Figure 2.34. View of curves DTA (a) and DTG (b) of dyed PVA-films: 1 – undyed, 2 – dyed by LIX, 3 – dyed by LX, 4 – dyed by LXI, 5 – dyed by LXII.

Comparing thermal stability of PVA samples, dyed by active dyes with formation of chemical bonds and by deactivated dyes , it may be observed that thermal stability of samples, dyed covalently, is a little higher than PVA containing the same dyes in inactive form. So, loss of mass for the sample, dyed by the colour LIX during 5 hours is 4,5%; LX – 6%; LXI – 6,7%; LXII – 10,5%.

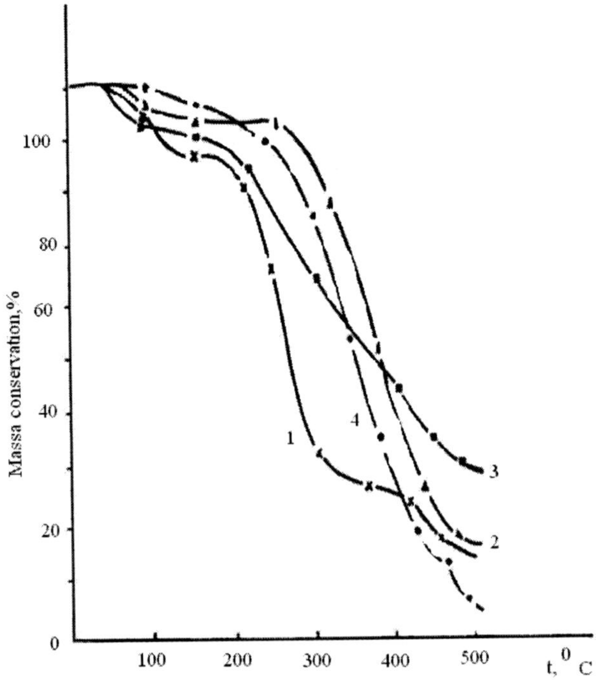

Figure 2.35. Thermal stability of covalently dyed PVA: 1 – undyed, 2 – dyed by LXIII, 3 – dyed by LXI, 4 – dyed by LIX.

Taking into account the fact that polymer materials must work for a long time in narrow temperature limits kinetics of thermal decomposition at thermal heating has been investigated. Data on kinetics of loss of mass at sample heating in the air at 200°C are given in Figure 2.36, from which it is seen that initial PVA (curve 1) displays the highest loss of mass. So, if undyed PVA loses about 15% of mass during 5 hours, then the one dyed by the colour LXII – 1,8%.

From Figure 2.37 it is seen that introduction of dyes increases thermal stability of PVA in the air, moreover it depends not only on the character of the found dye-polymer, but on chemical structure of the dye itself. So, the highest effect appears in phthalocyanine dyes, then follow azodyes and antrachinone dyes have the least effect. At the same time general tendency to improve stability to thermooxidative destruction for samples, containing covalently linked dyes, is displayed here.

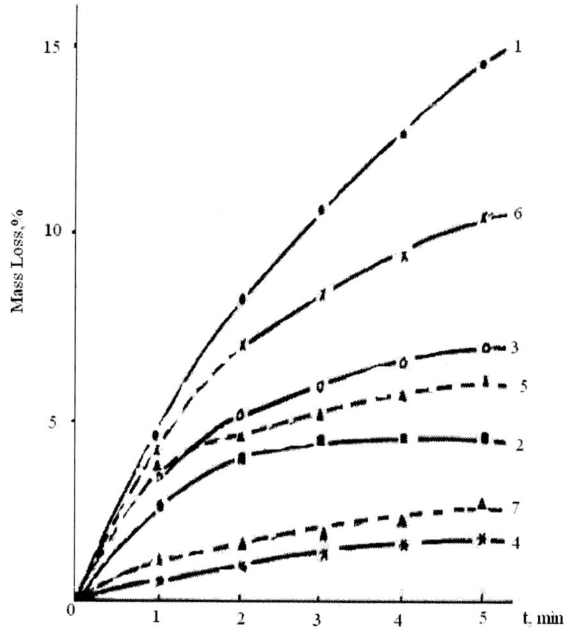

Figure 2.36. Kinetics of decomposition in the air of PVA-films at 200°C: 1 – undyed; 2 – dyed by LIX; 3 – dyed by LX; 4 – dyed by LXII; 5 – dyed by LXI; 6 – dyed by LXIII; 7 – dyed by LXI.

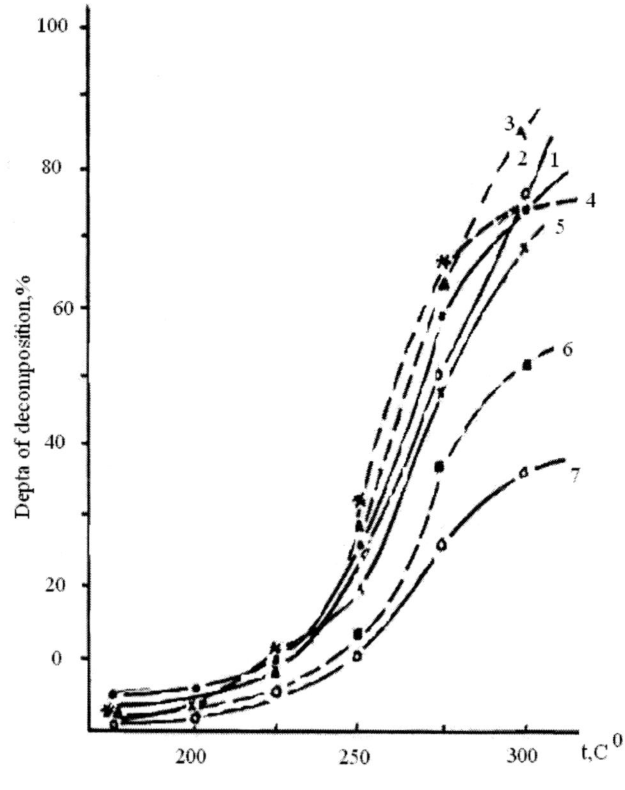

Figure 2.37. Decomposition of PVA in the air. 1 – undyed; 2 – dyed by LX; 3 – dyed by LXIII, 4 – dyed by LXII; 5 – dyed by LIX, 6 – dyed by LXIV, 7 – dyed by LXI.

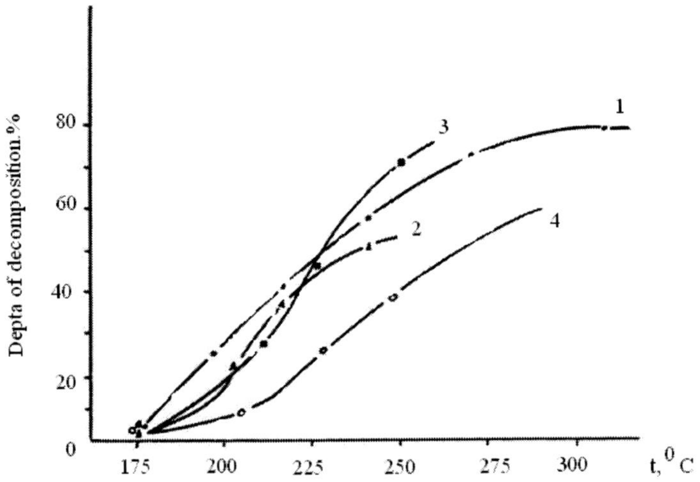

Figure 2.38. Thermal destruction of PVA-films in vacuum – initial (1) and dyed (2-4): 2 – LIX, 3 – LXI, 4 – LXIII at 175-250°C.

Curves of the loss of mass of dyed PVA samples at warming up in the air during an hour depending on temperature of warming up are given in Figure 2.38. They show that at temperatures of warming up to 225°C the loss of mass is not more than 10% but at further rise of temperature the loss mass increases.

Introduction of covalently linked dyes LIX, LXI, LXII reduces loss of PVA mass, moreover the highest effect is observed in phthalocyanine dye LXII, then in LIX and antrachinone dye LXI hardly influences the depth of polymer decomposition. Inactive dyes LX and LXII even slightly increase the depth of decomposition, exception is phthalocyanine dye LXIV, which shows stabilizing effect though much less, than covalently linked dye LXIII. So, depth of decomposition of undyed PVA during warming up at 300°C is 74%, while introduction of dye LXIV reduces it up to 51% and covalently linked dye LXIII – up to 35%. Oxidation processes of initial and dyed by active and deactivated dyes PVA are characterized by complex reactions, which distort the picture of decay.

A little different picture is observed during warming up in vacuum. Curves of loss of samples mass, warmed up during an hour, depending on temperature, are given in Figure 2.38. As it is seen the loss of mass of undyed sample increases almost linearly with the rise of temperature up to 250°C, while inhibitive effect up to 200-215°C is observed in dyed samples and this shows that oxygen is the first to react under oxidation at this temperature. At much higher temperatures depth of dyed samples decomposition increases very quickly, and for the dye LXI it even exceeds in initial PVA. Depth of decomposition is less in samples dyed by azodye LIX, especially phthalocyanine LXII. At temperature 250°C the rate of mass loss is 0,75g/min (Figure2.38, curve 2) and 0,5 g/min (Figure 2.38, curve 4) respectively.

When analyzing gaseous product of PVA decomposition in vacuum it was stated that main volatile products are H_2O was defined quantitatively by the reaction with calcium hydride and by gasochromatography determination of hydrogen, being released during the reaction:

$$CaH_2 + 2H_2O = Ca(OH)_2 + 2H_2$$

Table 26. Amount of water in % from the mass of the sample, released during warming up dyed PVA in vacuum during an hour

Temperature, °C	Undyed PVA	PVA, dyed by colours					
		LIX	LX	LXI	LXII	LXIII	LXIV
175	0,3	0	0,4	0	0	0	0
200	2,4	2,3	2,4	1,9	2,1	0	0,8
225	5,8	7,2	6,5	8,5	9,1	1,9	2,9
250	12,7	7,6	12,3	9,7	11,5	4,1	5,1
300	12,2	-	-	-	-	8,2	-

From the data of Table 26 it is seen that H_2O quantitatively being released at thermal decomposition of initial PVA and PVA dyed by inactive forms LIX and LXI, in being investigated temperature range, is characterized by close coefficients, what was to be expected, because oxichlortriazine dyes are not able to react with OH – PVA groups. Quantity of released water in PVA samples, covalently linked with dyes LIX, LXI, LXIII, at 250°C is much smaller than in initial PVA, moreover the least water is released from a sample dyed by phthalocyanine dyes of LXIII type. Substantive phthalocyanine dye LX behave somewhat differently than inactive forms of dyes LIX and LXI which is probably connected with the possibility of partial blocking by it OH-groups of PVA by means of complex-formation shown in works [73,79] for the series of direct dyes.

Results of EPR and viscosimetric investigations of dyed and undyed PVA-films are worth of attention.

So, before irradiation the dye LIX did not have paramagnetic properties. EPR signal in the form of singlet appears only after 4 hours of irradiation (Figure 2.39). This signal is stable, its intensity changes a little and in an hour after stopping irradiation. The signal disappears when irradiated dye is being dissolved in water.

Unlike LIX the dye LXIII displays paramagnetic properties and before irradiation. EPR – spectrum, shown in Figure 2.39, is a singlet.

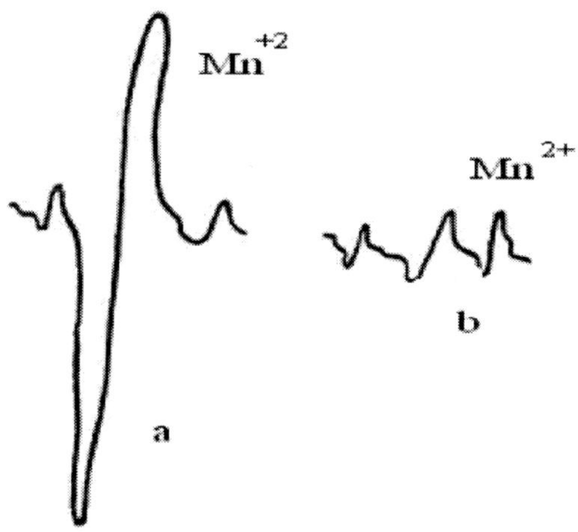

Figure 2.39. EPR – spectra of dyes LIX (a) and LXIII (b) after 4 hours ultra-violet irradiation.

Presence of conjugated π – bonds in the dye structure facilitates uniform distribution of π – cloud of unpaired electrons, that is why singlet does not change at excitation. Kinetic data of radicals accumulation in irradiated dyed PVA-films are given in Figure 2.40. From this figure it is seen that dyes LIX and LXI retard the rate of radicals formation in PVA to a greater extent than dyes LX and LXI. Such effect of increasing photo- and thermal stability of dyed PVA-films may be classified as partial structurization of polymer during thermal treatment and ultra-violet.

Presence of the processes of structurization in undyed film during the first period of irradiation may be observed while investigating viscosity of irradiated PVA-films, that proves some increase in intrinsic viscosity (see Figure 2.41). Such increase in intrinsic viscosity is not observed in dyed films. Probably PVA structurization after ultra-violet irradiation takes place by recombination radicals of polymer adjacent chains. The rate of radicals accumulation in dyed PVA-films is less (Figure 2.41) and at the same time probability of structurization by recombination PVA macroradicals decreases.

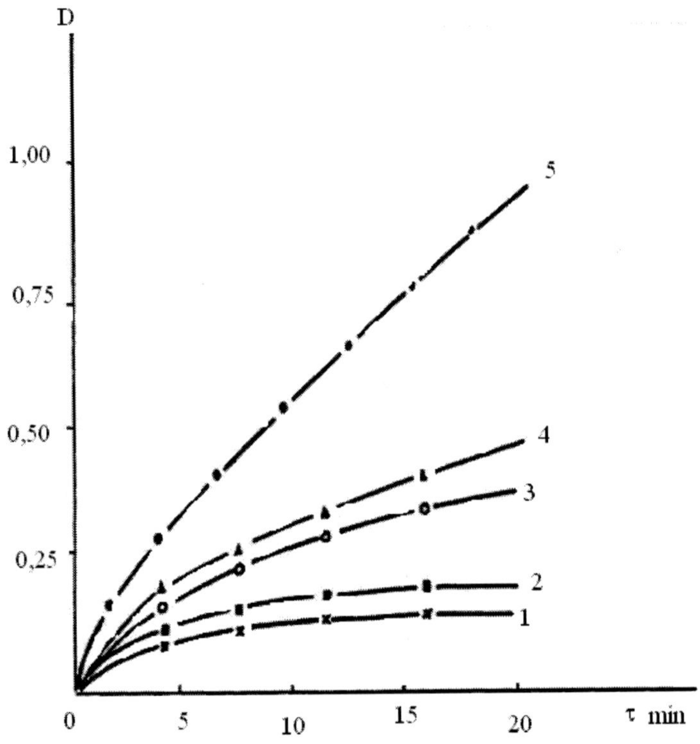

Figure 2.40. Kinetic of free radicals formation at ultra-violet irradiation: 1 – PVA covalently linked with LIX; 2 – PVA covalently dyed by LXI; 3 – PVA, dyed by LX; 4 – PVA, dyed by LXII; 5 – undyed PVA.

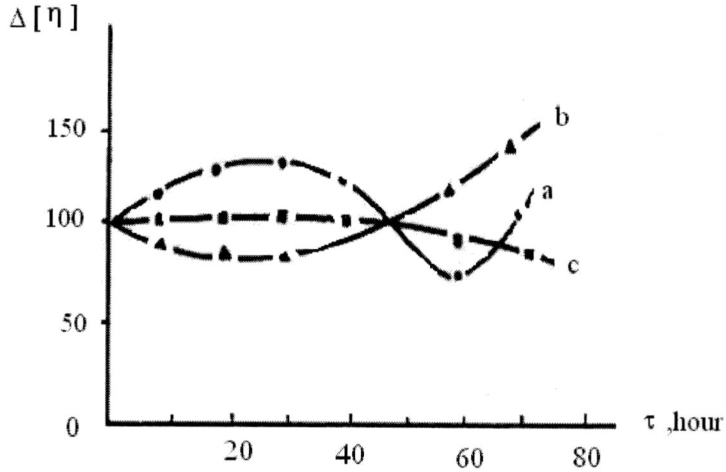

Figure 2.41. Kinetic of intrinsic viscosity change at ultra-violet irradiation of PVA: a) –undyed; b)-dyed I; c)-dyed II.

Thus, we may come to a conclusion that dyeing of polymers, containing free OH-groups, by active dyes considerably improves their thermal and photooxidative stability; this is connected with substitution of hydrogen labile atom by more volumetric molecule.

Chapter 3

STABILIZATION AND MODIFICATION OF POLYETHYLENETEREPHTHALATE

Polyethyleneterephthalate (PETP) is widely used in industry as fibres and films. PETP belongs to thermoplastic ethers and is a product of polycondensation of ethylene glycol and dimethylterephthalic acids [186].

Polyethyleneterephthalate is produced by two stages: interesterification of dimethylterephthalate by ethylene glycol takes place at the first stage:

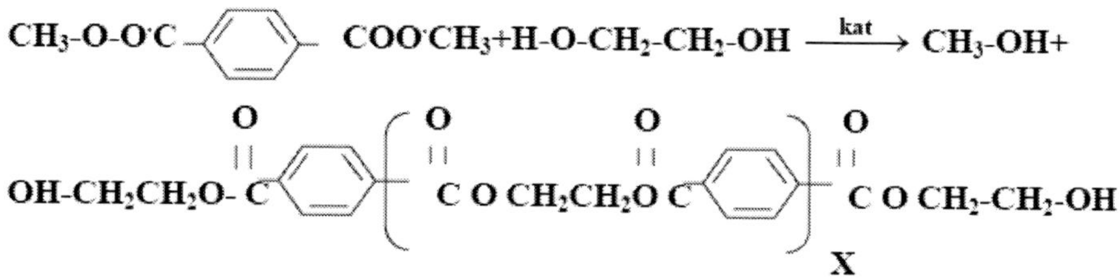

x =2-4;

polycondensation, which flows with ethylene glycol splitting out, distilled off under vacuum [187] takes place at the seconds stage.

PETP molecule has two characteristic features. First – it is regularity of phenylene groups arrangement, which are linked with each other by ester groups $-\overset{O}{\overset{\|}{C}}-O-$

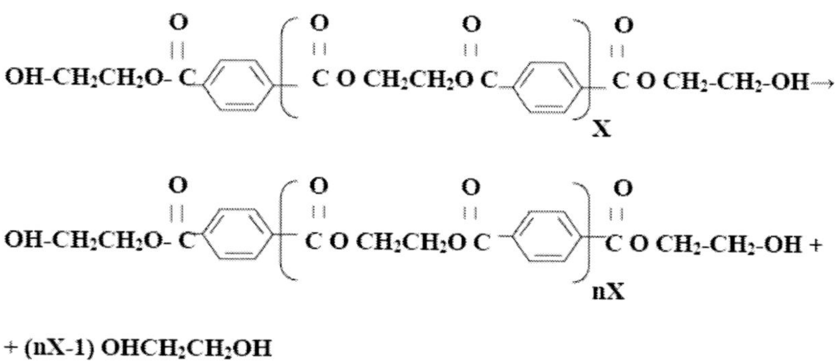

Another characteristic feature of polyethyleneterephthalate is almost plane configuration of chains and presence of two centres of symmetry on each repeat unit [188]. These two features cause the ability of PETP to crystallization [189]. Crystallization of PETP flows with formation of spherulite structure leading to polymer turbidity.

These features stipulate properties of PETP-fibre. PETP-fibre is classified as strong, resilient, elastic fibre, having low creep. Fibre strength depends on molecular weight of the polymer, its dispersion, on the processes of shaping and drafting of the fibre. Permissible operating temperature of use is 120-130 ^{0}C. Important advantage of PETP filaments is their high resistance to tensile deformation. PETP fibre is thermoplastic – it softens, begins to glue and loses its form at intense heating. These changes take place in temperature range of 180-200 ^{0}C[190]. Owing to thermoplasticity strength characteristics decrese and breaking elongation increases.

Polyester fibres, specifically PETP, possess high thermal stability and sufficient resistance to light effect and atmospheric action. However, effect of high temperatures, to which PETP is subjected in the process of synthesis and processing into products, may cause thermal destruction of polymer and in the absence of oxygen. Decrease of molecular mass, increase of end groups quantity, extraction of volatile products, the main of which are terephthalic acid, acetaldehyde and carbon oxide are observed at PETP thermal destruction. Besides these substances, anhydride groups, benzoic acid, p-acetyl-benzoic acid, ketones, acetals are identified in the compound mixture of the products of thermal decay [34].

As a result of thermal destruction, owing to formation of volatile products polymer mass losses take place. Mass losses at 310 ^{0}C are 1%. The main stage of decay (up to 85%) ends at 487 ^{0}C and at 610 ^{0}C polymer completely burns out [191].

It should be noted that the results of thermodestruction investigations made by different authors are different which is caused by different conditions of experiment carrying out, specifically, atmospheres of different inert gasses have been used for investigation of thermal destruction.

The process of PETP thermal decay in argon is similar to the process taking place in nitrogen atmosphere, while in the air mechanism of destruction is different. It is shown in the work [192] that in the atmosphere of inert gas only one stage of decay takes place and all kinetic parameters of decay (temperature of maximum mass losses, energy of activation, rate of destruction and so on) grow with the increase of molecular weight. But in the atmosphere of air two stages of PETP-fibre destruction are observed. At the first stage kinetic parameters of decay decrease with the growth of temperature and polymer molecular mass. These changes are expressed more brightly at the second stage of the decay process.

Surrounding atmosphere influences the formation of either decay product in the process of destruction. So in the work [193] it has been found that in nitrogen atmosphere at isothermal heating at 160 ^{0}C mainly benzoic acid and esters are formed. Since oxidative processes are absent at thermal destruction in nitrogen atmosphere then one may come to a conclusion that benzoic acid and esters are the products of thermal decay of PETP fibre. At thermal destruction of PETP-fibre thermal decay plays a very important role, especially at the beginning of the destruction process.

Results of above mentioned works, where the effect on thermal destruction of alcoholic residues, constituting polyester are very important. Polytrimethylterephthalate has been synthesized in the work [192], and polybutylene terephthaeate – in the work [193]. Thermal stability of PETP and polymethylterephthalate does not differ very much, and thermal stability of polybutylene terephthalate is less, than that of PETP and polymethylterephthalate.

On the basis of the results of investigations [192, 193] it was proved in the work [194] that thermal stability of polyester will depend on one of the initial reagents (alcohol) taken for polymer synthesis.

Decrease of polymer molecular mass takes place during the process of destruction. In the work [195] there has been shown linear behaviour of the dependence of lg(da/dt) on t (time), which speaks about the fact that the quantity of broken bonds α at any moment of time is defined by the equation $\alpha=1-e^{kt}$, hence, decrease of molecular mass at PETP destruction happens as a results of bonds decay according to the law of randomness. This conclusion is proved by PETP destruction in nitrogen flow at 280°C which does not lead to the change of the character of molecular – mass distribution (MMD) [196].

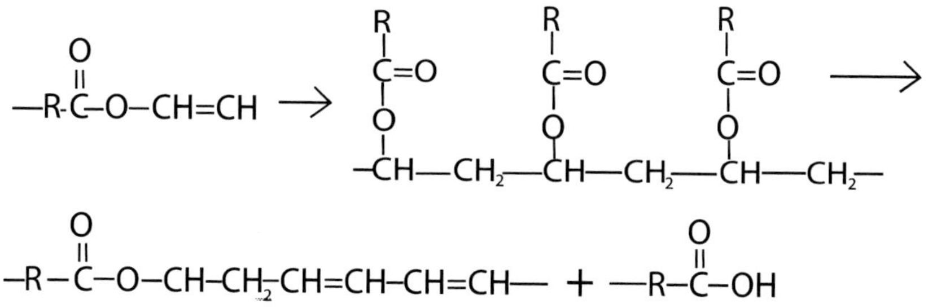

During thermal destruction colourless PETP-fibre is dyed into cream-colour, then it turns yellow and further becomes brown. It was shown earlier that darkening of decomposed PETP is connected with polymer product dyeing and is not caused by the presence of dyed low-molecular compounds. Vinyl ester groups in decomposed PETP even at temperatures up to 300 ^{0}C are able to come into reactions of polymerization [197].

Besides there appear branched chains, containing sections of polyvinylbenzoate type, which may again decompose through ester bonds with hydrogen atom break in β-position regarding ester bond end forming and carboxyl groups: R-C_6H_4.

So, there appear systems of conjugated double bonds which cause polymer dyeing.

It should be noted that some authors [198, 199] connected polymer dyeing at thermal destruction with aldehyde polymerization:

$$nCH_3CHO\ CH_3(CH{=}CH)_{n-1}+(n-1)H_2O \quad \rightarrow$$

There has been also voiced a supposition about simultaneous proceeding of both processes [200].

In the work [201] it was supposed that chemical behaviour of PETP thermal decomposition both in inert medium and in atmosphere is defined by the mechanism assuming formation and decay of intermediate cyclic compounds, catalyzing the process.

Prokopchuk N.P. [199] with his research workers suggested the scheme of radical mechanism of PETP thermal destruction with initiation by the way of C-H bond break:

$$-C_6H_4\text{-}COO\text{-}CH\text{-}CH_2\text{-}OOC\text{-}C_6H_4\text{-} \rightarrow -C_6H_4\text{-}COOCH{=}CH_2 + \text{-}OOC\text{-}C_6H_4\text{-}$$

This scheme explains the formation of carboxyl groups, acetaldehyde, benzaldehyde and also considerable quantity of CO_2 during PETP thermal destruction.

Water, appearing in the system during forming polyene structures from the product of PETP thermal destruction – acetaldehyde, causes hydrolysis of ester bonds, increasing the rate of carboxyl groups formation. Besides, there takes place decrease of polymer relative viscosity [202].

On the basis of mass – spectroscopic method there was made a supposition that the break takes place mainly in carboxyl groups in three states [203]:

$-CH_2-CH_2-/-CO-/-O-/-C_6H_4-$

Mass – spectroscopic investigations are proved by pyrograms [204].

Kinetic parameters of thermal destruction were defined by DTA method in the work [205]. Obtained data correlate well with the results received from the curves of kinetics of carboxyl groups concentration increase.

Presence of CH_2 – unit in aliphatic groups leads to the decrease of thermal stability. Attack during thermal destruction is directed onto α-CH_2 – group, as this group in diethyleneoxide fragment differs by particular vulnerability [206].

Time of temperature effect on the sample influences the process of PETP destruction. During short effect sharp decrease of polymer specific viscosity takes place while polymer mass does not change. Drop of viscosity value slows down during the increase of temperature effect time and polymer mass loss sharply increases. Viscosity change is caused by thermal destruction of macromolecule, and mass change – by the destruction of chain end groups [207].

Thermal destruction begins at lower temperatures in the air than in the inert medium. This occurs because of the presence of oxygen [208].

Rate of thermal destruction is 10 times higher in the presence of oxygen than in the inert medium [209]. At relatively low temperatures (up to 140 ^{0}C) PETP fibre is quite resistant to thermooxidation, but at the temperatures above 220-250 ^{0}C oxidation processes flow with considerable rate [210]. Thermal destruction in the presence of oxygen goes under the complex mechanism including thermooxidation.

During thermooxidative destruction of PETP-fibre at 280 ^{0}C [211] there were found the same products of decomposition as during thermal destruction (in atmosphere of helium) but in much larger quantities.

A great number of investigations are devoted to the study of thermooxidative destruction by the method of differential thermal analysis (DTA) [212, 213]. During DTA curves analysis it has been found out that resistance to PETP thermooxidation decreases with the increase of its molecular mass [212].

All temperature transitions characteristic for thermooxidative destruction were well observed on DTA curves. On DTA curves it is clearly seen that exothermic peak of oxidation is being observed in the presence of oxygen in the melting region. Besides, DTA curves show that PETP thermooxidation is slowing down as the temperature is approaching melting point, and this is connected with the softening of the sample and further decrease of its surface. DTA method was used to define kinetic parameters of thermooxidative destruction. Calculated parameters correlate quite well with the results obtained from the curves of kinetics of carboxyl groups concentration increase at isothermal oxidation of PETP [213].

With the help of DTA method effect of interesterification and polycondensation catalysts on PETP stability at thermooxidation [214] has been studied. Thermooxidative destruction is accelerated by the presence of metal catalysts.

Labile structures initiating polymer decay [215] are formed during the process of thermooxidation in the air. Thermooxidation rate is defined by the rate of oxygen diffusion into polymer. Constant of destruction rate in the air compared with inert medium increases, and activation energy decreases [216]. However, in some cases active energy increases; this is connected with the contribution of physical phenomena of heat and mass transition together with chemical processes into the total kinetics of destruction.

It is supposed that PETP decay during oxidation goes mainly along ester bonds by their hydrolysis with water, formed at hydroperoxide decay [217].

In the work [218] there have been found hydroperoxide sections being formed during oxidative destruction and responsible for thermal stability of PETP. It is shown in the work that prediction of antioxidative behavious of PETP directly depends on the quantity of hydroperoxide radicals being formed.

Depolymerization of polymer macromolecules runs as a result of sample destruction, moreover the greatest rate of depolymerization is observed at the temperature 533-543 K and pressure 9,0-11,0 MPa [219]. Products of depolymerization were analysed with the help of gas-liquid and gas chromatography.

Absorption of considerable amounts of oxygen at high temperature oxidation of PETP shows that oxygen takes part not only in initiation but in further stages of radical-chain process [34].

So, all works on investigation of the mechanism of thermal and thermooxidative destruction suppose that these processes have radical nature and run by the way of formation and decay of peroxide radicals and hydroperoxides. Together with oxidative decay of aliphatic part, as a result of which polymer chain decomposes with volatile products liberation and formation of new end groups, there are also changes in aromatic part-jointing of polymer takes place.

3.1. Photo- and Photooxidative Destruction of PETP

PETP is quite resistant to ultra-violet irradiation but at long action of sun-light destruction becomes visible. Light stability of PETP depends on irradiation conditions, and irradiation intensity and wave length are the most important factors here. Length is also important and so energy of light waves being absorbed by the fibres during insolation is important too. Thus, effect of irradiation and its intensity are defined by the energy of light waves.

PETP destruction under atmospheric conditions, as a result of photochemical reaction, flows mainly under the action of ultra-violet rays with λ=300-330 nm [220]. Energy of these waves is about 1% of the total energy of the sunlight. Authors connect this fact with the presence of absorption maximum of PETP itself in this region of the spectrum.

Considerable changes take place in PETP structure at ultra-violet irradiation. So M.S. Kuligin with his research workers found increase of absorption bonds intensity in the region 1620 cm^{-1} in infrared spectra of irradiated PETP, characterizing the presence of double bonds

–C=C-, that is displaying formation of unsaturated groups [221]. They consider that light absorption with λ below 300 nm is caused by groups C=O or

$$\begin{array}{c} -C-O- \\ \| \\ O- \end{array}$$

[158].

And those groups, which do not destruct themselves, may absorb light energy, causing molecule destruction. Excitation in the region 340 nm relates to π*→n transition, and fluorescence in the region 380-460 nm relates to S_1 (n- π*) state and interaction of C=O groups with π – electrons of phenylene. Fluorescent properties of PETP appear in the presence of abnormal units which are formed as a result of stilbene structure impurities taking part in polycondensation and being present in monomer [222]. It is supposed that absorbed energy may quickly migrate to other parts of molecule [10] in PETP as in other polymers.

Ester groups contained in PETP are subjected to the greatest transformations during irradiation. In the work [221] it is shown that ester groups in PETP are exposed to the greatest destruction under the action of ultra-violet irradiation as a result of which carboxyl and hydroxyl groups are being formed. And this explains the increase of absorption bands intensity in infrared spectra of PETP irradiated by ultraviolet light, these bands being related to these groups.

Direction and rate of changes depend on wave length of incident light, intensity of radiation and PETP structure. Investigations [115] on the study of PETP fluorescence at excitation by light with wave length λ=340 nm show that excitation spectra depend on relation of emission intensity at λ_{max}370 and 390 nm and on film thickness.

According to the data, given in the works [223-224], it is seen that during PETP irradiation by both polychromatic and monochromatic light with different wave length distant ultra violet light up to 330 nm has the greatest destroying effect.

Arc carbon lamps, used in apparatuses of "Fedomer" type [225], xenon gaseons-disharge lamps of high pressure, used in apparatuses of "Xenotest" type [226], tubular luminescent erythritol lamps [227], mercury-quartz lamps [228] and other devices [229], used in some countries while carrying out tests on light-fastnes of point, were proposed for light-fastness of paint test.

Products of photodestruction are defined, as a rule, by the methods of EPR, infrared spectroscopy, gas- liquid chromatography. Carbon oxide and dioxide, hydrogen, methane, water, benzene, formaldehyde were identified by these methods.

Sharp change in supermolecular structure [230] is observed during artifical irradiation.

Great disagreement of results obtained while using different light sources may be explained, on the one hand, by differences in spectral composition of radiation and, on the other hand, by difference in temperature and humidity of the sample since PETP reacting with water is subjected to destruction (hydrolysis takes place). The process of hydrolysis is catalyzed by bases and acids. It is the presence of acids and bases, contaminating the atmosphere that explains acceleration of PETP photodestruction while studying the effect of lightning on the polymer [231, 232].

Direct photolysis or, as it is sometimes called photochemical destruction, leads to the break of chemical bonds, absorbed by light, in macromolecule. Photochemical transformations may be caused by radiation in the region 100-400 nm. This energy is quite enough to break the bond C-C or C=O and also C-H [154]. Hence, in order for the destruction to flow according to the mechanism of direct photolysis, PETP macromolecule must absorb light with the wave length not higher than 400 nm. However, absorption of one quantum of light not always leads to the formation of one particle of photochemical transformation.

This phenomenon may be caused by a number of reasons. Active molecule may not at all decay chemically even if absorbed quantum has much more energy than the energy of dissociation of the strongest molecule. This is probably connected with the fact that energy absorbed by particular sector of molecule may be distributed along different bonds in molecule.

The second reason is that initially formed radicals may recombinate faster, than react with other substances and so excess of energy will be transformed into kinetic energy.

During insolation in natural conditions there takes place weakening of molecular bonds in amorphous section of polymer.

Insolation in natural conditions and under artifical light source shows that in both cases change of the value of PETP molecular weight happens less intensively compared with physico-mechanical properties. Evidently, the fact, that physico-chemical properties are defined not only by chemical structure, but by complex formations of supermolecular structure, is characteristics for PETP-fibre.

A great number of investigations were devoted to the problem of oxygen effect on photodestruction of polyesters. Many researchers consider that the rate of photodestruction depends on the presence of oxygen [233-235].

Plants, allowing to carry out irradiation both in aerial and airless media [233] are designed for studying oxygen effect on the rate of photodestruction.

During irradiation in the absence of oxygen the main effects are joints between polymer chains [234]. Break of the chain and increase of fluorescence are observed in the presence of oxygen, but there are no lateral joints between chains and only weak decolouration is observed. Another great difference between photolysis both in the presence and absence of oxygen is in the quantity of released CO_2 [235].

The important breaking factor for PETP is humidity of air and soil [236]. Besides, the process of photodestruction is catalyzed by the presence of copper saults [237].

Not excluding the possibility of more complex mechanism of PETP photodestruction and photooxidation it is supposed that photodestruction of the given polymer takes place according to the Norrish reaction of I and II type [2].

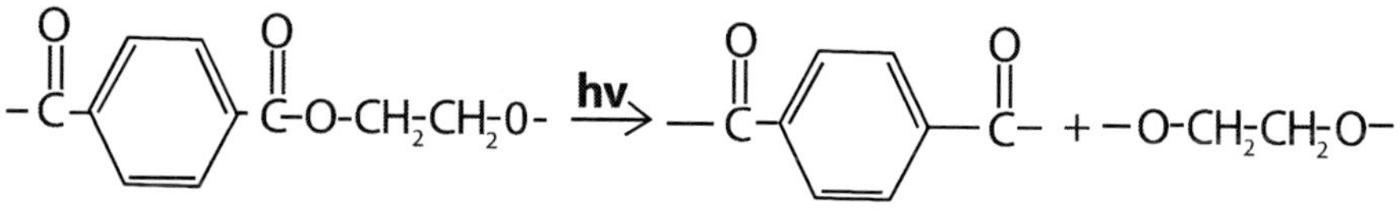

Norrisch reaction of I type is the break of the main chain of polymer macromolecule under the action of ultra-violet light quanta.

According to the Norrish reaction of the II type there occurs intramolecular separation of the hydrogen atom and formation of the radical with its following decay [238].

Increase of irradiation time causes decline of the properties of PETP-fibre; namely elogation at rupture and strength. Decrease of strength characteristics and increase of terephthalic acid content is proportional to irradiation time; hence, according to the quantity of terephthalic acid one may judge about degree of photodestruction [239]. However, terephthalic acid, formed as a result of PETP photooxidative destruction, to a considerable extent absorbs ultra-violet rays, protecting deep layers of the fibre and inhibits the process of their photooxidative destruction [240].

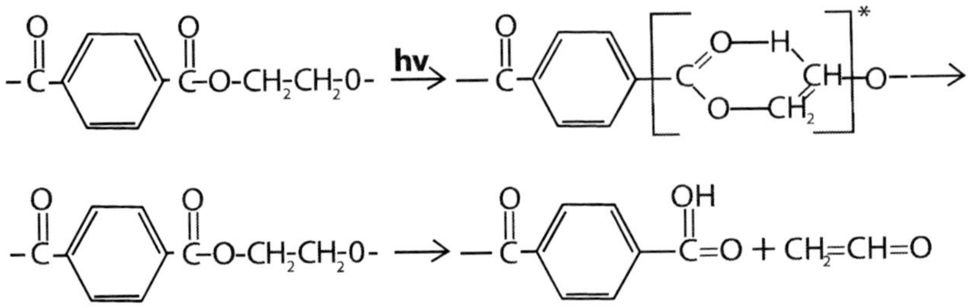

Many reactions of photosensibilized oxidation run with oxygen, taking part in it, and being in electron-excited singlet state. Singlet oxygen may form products of oxidation with the yield being close to theoretical [241].

When it was proved that singlet oxygen takes part in reactions of photosensibilized oxidation Trossol and Winslow have suggested new mechanism of photooxidative destruction of polyesters including 4 stages [242]:

(1) light absorption by carbonyl groups;
(2) photodecomposition in the reaction of the II type according to Norrish, proceeding with the participation of excited state n→π* carbonyl groups;
(3) formation of singlet oxygen in reactions of suppression of triplet state n→ π* carbonyl groups;
(4) reaction of singlet oxygen with vinyl groups formed according to reaction of the II type according to Norrish.

This mechanism includes only initial stages of reaction. Reactions following the formation of peroxide compounds are supposed to be the same as in more widespread kinetic schemes describing the course of radical-chain processes. Hence, one may come to a conclusion that during polyesters oxidation there proceed two processes at one and the same time:

(1) leading to the formation of peroxide compounds (radical-chain);
(2) proceeding with singlet oxygen participation.

Studying mechanism of photo- and photooxidative destruction Ibishkovitch and others [243] have shown that during PETP photodestruction intrinsic viscosity decreases, carboxyl groups contain increases, sample density rises, optical density increases at 373 and 632 cm"1

Empirical equations, describing kinetics of the change of PETP characteristics have been found in the same work:

$$\eta = o{,}741 \bullet t^{0,1};\ [\text{-COOH}] = 0{,}05 \bullet t^{0,8};\ P = 1400 + 42\alpha \bullet 10093t;\ D = 3 + 0{,}263t^{0,263}$$

Change of molecular weight distribution of the polymer [244] occurs at the increase of the sample optical density and exposure dose, that is why corresponding corrections to the molecular-weight distribution [245] should be introduced into the results of measurements. In order to avoid the necessity of introducing corrections to molecular-weight distribution the tests are carried out on thin films (3-5 nm) with optical density not higher than 0,35. Exposure doses must be low causing no more than one break counting one initial macromolecule [246].

The rate of any photoreaction equals $v=\Phi I_{abs}$, where Φ – quantum yield; I_{abs} – amount of light, absorved in unit time.

Thus, the most important factors, on which PETP destruction depends, are intensity of irradiation, wave length and presence of oxygen. In order for destruction to proceed according to mechanism of direct photolysis PETP must absorb the light with wave-length not higher than 340 nm, moreover the process of destruction must proceed in the atmosphere of inert gas or in vacuum. In this case the main effects are joints between polymer chains.

Photoxidative destruction of PETP will proceed in the presence of oxygen. Breaks of macromolecules chains will take place in polymer, but lateral joints between chains are not observed. Two processes proceed during photooxidation at the same time: first – radical-chain, second – with singlet oxygen taking part in it.

3.2. Methods of PETP Modification

Modification of physical and chemical structure of polyster fibres allows to broaden greatly the assortment of these fibres, to give new valuable properties, which, in some cases, are not characteristic for the fibre from homopolymer.

There may be different directions of modifying action: thermal stabilization, light stabilization, plasticization, dyeing.

There are two methods for stabilization of polymers:

(1) introduction of special additives-stabilizers;
(2) modification by physicl and chemical methods.

The example of the second method is: giving fire-retardant properties to PETP – antipyrine is introduced by the way of additive mixing with polymer [247] plasma modification is used with the aim of improving adhesion [248].

The first method is more perspective, as stabilizers may inhibit destruction reaction; may directly influence the mechanism of destruction with the purpose of decreasing undesirable products yield and so on. It is supposed that stabilizers may act by means of: 1) blocking of active centres (weak bonds); 2) filtration of ultra-violet radiation; 3) breaking of peroxides; 4) interaction with free radicals; 5) suppression of excited states.

Connection of modifier structure with the effect of its action and also interconnection of polymer structure with the structure of either additive are not finally revealed up to now. For many modifiers mechanism of their action is not known at all, though their efficiency is known.

While selecting modifier it is necessary to consider compatibility of the additive with polymer, volatility and extraction, dyeability, toxicity, odour, economical efficiency, effect on technological mode of processing, service properties of polymer materials.

One of the methods of introducing additives into PETP is modification at the stage of synthesis. To increase strength and resistance to thermodestruction of PETP filament it is necessary for it to have minimal content of diethylene glycol, carboxyl groups and high molecular mass [249]. Synthesis of PETP on the basis of ethylene glycol and dimethyl terephthalate may be intensified by introducing oligomers of polyethylene oxide with end epoxy group [250] at the stage of interesterification and polycondensation.

Addition of low-molecular monoepoxides at the beginning and at the end of PETP synthesis allows to increase PETP hydrolysis resistance, thereby to improve physico-chemical properties of the fibre, to obtain polyester fibre with high degree of polymerization, to increase greatly thermal stability of polyester fibres.

Thermal stability of PETP may be achieved by decreasing end carboxyl groups up to ≤ 15 g-ekv/10^6 at the expense of interaction of compounds 1 and 2 with end – COOH groups of polymer:

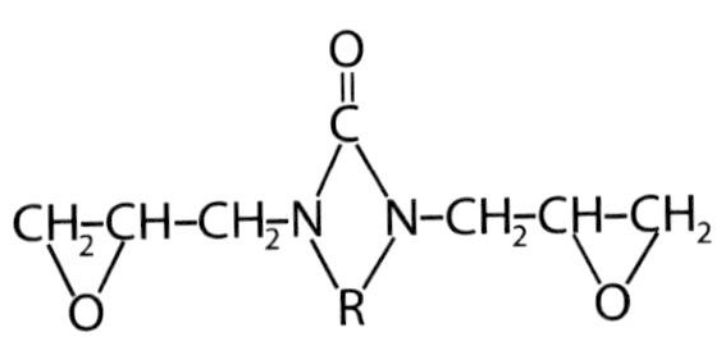

(1)

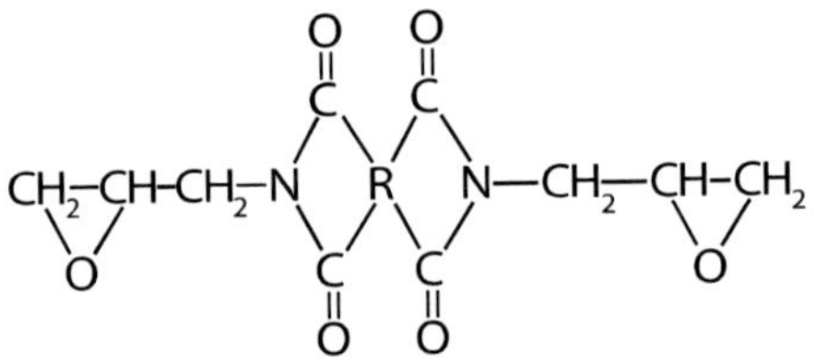

(2)

It is noted here that modified fibres conserve up to 81% of strength at thermal treatment up to 290 ^{0}C, whereas this value in unmodified fibre is 60% [251-252].

PETP with increased resistance to thermal destruction [253] is obtained in the presence of diglycol phosphate, igranox 1222 and trimethylborate.

Effective stabilizers of PETP are derivatives of phosphoric acid [254, 255], borites, containing naphthyl and p – hydroxyl phenyl groups [256], zink propyl diphosphat, amide of salicylic acid, derivative of pyrimidine and 4,6 ditredbutylene phenol [255], compounds containing phosphorus [257].

Sulphide derivatives of polychlorperidines of general formula ASR, where A-C_6Cl_6, R-H, CH_2CH_2O, $CH_2C(Cl)HCH_3$ [258] are suggested as thermostabilizing and structure-forming additive for PETP.

Important factor is the amount of introducing stabilizer. High stabilizing efficiency is achieved by introducing stabilizer in the quantity of 0,05% of the polymer mass [259].

Effect of thermostabilizers on the polymer properties was studied by different physico-chemical methods. For example, in the work [260] method of DSS (differential spectroscopy) was used to define the effect of polyester-imide on thermo-physical properties of PETP. By this method it was found out that polyester-imide reduces PETP ability to crystallization. Methods of thermogravimetric analysis (TGA) and infrared spectroscopy in the nitrogen atmosphere were used in the work [261] to define thermal stability of the mixture of PETP and polyamide with the additive - modifier - polyethylene. It has been found that introduction of the additive decreases activation energy which positively tells on the ability of PETP to thermal destruction.

The task of light stabilization is the increase of light stability of polymer materials in the process of photooxidative ageing. Different light stabilizers are used to execute this task.

Phototransformation of polymers is, in general case, complex multistage process. Ultra-violet absorbers, ultra-violet shields, suppressors of excited states, inhibitors-acceptors of radicals and inhibitors, destroying products, in which photobranching takes place, are used to decrease the rate of phototransformation. Great importance acquires determination of initial photochemical processes to solve the problems of photodestruction of dyed fibrous materials. Primary photochemical reactions are accompanied by dark processes. Products being formed during dark and photochemical reactions may be subjected to further transformations. Hence it follows that influencing directly both photochemical stages and dark processes one may protect polymer from light ageing.

Such characteristics as compatibility with polymer, volatility, chemical stability under conditions of processing and usage become very important during selection of light-stabilizer.

It should be noted that efficiency of additives action depends on the testing conditions (intensity and spectral composition of the irradiation source, humidity, concentration of additives and so on). This is connected with the fact that stabilizers may at the same time act according to several mechanisms, but only one of them is realized depending on testing conditions.

Japanese development engineers have shown that fibres, reflecting ultra-violet rays [262] are obtained during impregnation of polymer, containing finely grinded powder of ZnO_2, MgO, $BaSO_4$ or $MgCO_3$, by butadiene and chloroprene solution. Another method of PETP light stabilization is introduction of 0,1-2% of finely dispersed inorganic particles [average diameter $\leq$ 0,01-0,05 nm]. Fibres being obtained have improved light stability [263], [264].

Triphenylphosphate, igranox, diglycol phosphate, trimethyl borate [265], benzotriazole [266] are used as light stabilizers. Fibres modified by such substances exceed goods from compositions produced by traditional method to a far greater extent as to their resistance to ultra-violet irradiation.

In the work [267] ultra-violet absorbers, black and mixture of TiO_2 with $BaSO_4$ were used as light stabilizers. Obtained PETP samples were subjected to the action of light-weather during 1100 hours. Results of the light effect on modified PETP-films were analysed by ultra-violet spectroscopy, titration, chromatography and intrared spectroscopy. By these methods it has been found out that among above-mentioned stabilizers the most effective is ultra-violet absorber.

Good light stabilizers, used in the industry, are o-oxybezophenones [268]. Probably mechanism of their light stabilizing action is in the fast tautomerism of excited states of molecule

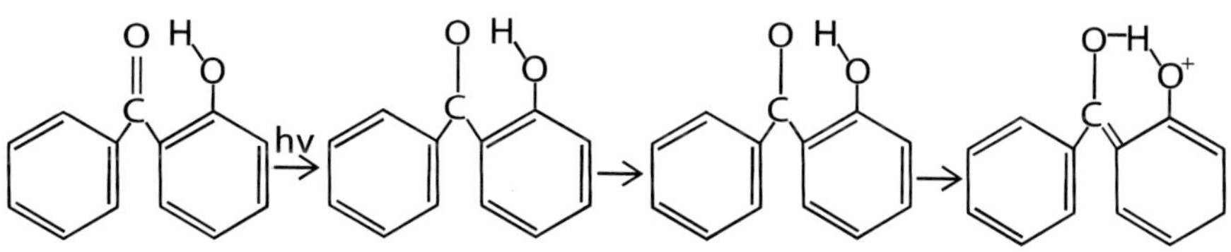

Hydrogen atom of OH group, situated in ortho-position, forms hydrogen bond with oxygen atoms of carbonyl group, thereby creating favourable conditions for photoenolization or proton transfer. Stabilizing action of o-oxybenzophenones increases because of this fact. The more labile is H-O bond in a molecule, the higher is the efficiency of their action.

Authors of the work [269], while studying mechanism of the action of benzophenones derivatives, proved the well-known conclusion [10] that light stabilizing action of these compounds is caused by fast, reversible phototransfer of the proton, taking part in formation of intra-molecular hydrogen bond but not suppesing excited states of polymer.

Analysis of DTA curves has shown that the main process of crystallization in the presence of stabilizer proceeds at much lower temperature. This is explained by nucleus activity of stabilizer and filling interspherulite space by it, preventing the growth of spherulites. It has been also obserbed on DTA curves that melting of stabilized PETP is observed at much lower temperature compared with unstabilized one, which is the result of less perfection of fine spherulite structure of stabilized PETP [270] and plasticizing action of the additive too.

Nonuniformity of stabilizer distribution in some cases may facilitate the increase of stabilization efficiency [58]. At nonuniform distribution, connected with crystallinity, lowmolecular stabilizers obtained on the basis of the same monomer as protected polymer [74] get definite advantage.

Thus, depending on the nature of the polymer and its structure, on the properties and mechanism of stabilizer action, high-molecular stabilizers may protect better or weaker than lowmolecular ones. However, advantage of highmolecular stabilizers becomes undoubted on exposure to the factors causing volatilization or extraction at the very beginning of extraction. Advantage of highmolecular stabilizers is displayed more brightly in such polymers which constantly or periodically must be in contact with water or organic liquids.

Use of lowmolecular stabilizers has essential disadvantages: they do not provide duration of protective effect under service conditions, since they are quickly consumed at ultra-violet irradiation, washed-out from the polymer and are lightly volatile. In this connection there arises the question about possible use of highmolecular stabilizers.

Lately, copolymers have been used as stabilizers [271]. In the work [272] copolymer on the basis of secondary kapron and polyethylentherephthalate is used for PETP stabilization. As a result impact elasticity rises by 1,5-2 times and deformation – strength properties by 40-70%.

According to technological reasons chemical additives in PETP are introduced mainly into the melt during the process of goods production. In some cases this is the only way for obtaining stabilized and dyed fibres.

3.3. Modification of PETP by Additives of Polyfunctional Action

As a rule, stabilizing additives are colourless [76]. However, in some case it is preferable to use dyes as stabilizers, especially if it is necessary to obtain dyed polymers [32, 77, 78, 84].

One of the most rational methods of PETP modification may be introduction of modifiers into polymer mass. At usual surface dyeing based on diffusive process of dye penetration into fibre inverse process is realized one way or another – that is yield of the dye from polymer matrix, which causes reduction of intensity of product dyeing. Besides, due to non-uniformity of fibres (presence of crystalline and amorphous regions) the dye is non-informly distributed in them, producing uneven painting and making colourity characteristics worse. It should be also noted that the process of bath dyeing is not economic because of incomplete bath extraction and necessity of carrying out additional finishing operations.

Coincidence of polymers synthesis or forming fibres from them with dyeing is progressive method in technological and economic respects by which fibres in complete marketable state may be produced directly at the chemical fibres factory.

It should be noted that in connection with strict measures on environmental protection stock dyeing, as practically wasteless production, becomes the most important among other methods. Besides, due to the shortage of service water there is no longer any necessity for sewage treatment of textile-finishing production from different impurities (dye, surface-active substances (SAS), salts and so on).

The most important fact is that dyes for stock dyeing in overwhelming majority provide colours with high stability indexes to all physico-chemical effects which are often unachievable during usual dyeing methods [273]. Dyes used for stock dyeing not only effect negatively on the fibre strength, but, in some cases, they are able to protect it from thermooxidative and photochemical destruction.

The process of obtaining and forming thermoplastics polymers is carried out at high temperatures, achieving 300 ^{0}C, which, in its turn, sets a strict limit on selection of dyes and restricts the assortment of the latter, as the majority of used dyes for PETP fibres do not bear long action of high temperature and agressive melt medium.

That is why for dyeing PETP-fibres in the process of obtaining or forming polymer there may be used only those dyes which meet definite requirements, namely:

(1) to be resistant to high temperatures (250-300 ^{0}C) throughout the process of obtaining from the melt, to be resistant to agressive melt medium, chemical action both on the part of separate components, being a part of the melt, and products of their chemical decay, and on the part of physical action during polymer processing;
(2) to solve and combine with polymer quite well, and in the case of pigment – to disperse well;
(3) to give the fibre paint stable to physico-chemical action.

The dye, being introduced into reaction mass at the stage of polymer synthesis when the additive undergoes monomers action, which, as a rule, are more reactive, than polymer, is under much harder conditions. Dyeing composition may be introduced at any stage of polymer synthesis in the form of powder, suspension and paste. Dyeing of granules is carried out by mixing with pigments powder in dry or wet form, in addition some part of the dye comes to waste waters which causes economic problems. This disadvantage may be eliminated if the dye will be introduced directly into prepared polymer mass.

Dyes, used for thermoplastic polymers, may be divided into two groups:

(1) soluble in spinning melt;
(2) high-dispersed pigments, which must be uniformly distributed in polymer melt and provide stable dispersions.

Both groups have their own advantages and disadvantages. In the first case great demands are made to dyes solubility, in the second – to the degree of dye particles dispersion.

Lately, a great number of new dyes has been synthesized, among them – azo dyes, water – soluble anthraquinone esters [274], derivatives of phthalocyanide [275], derivatives of piperidines [124], dyes of perylene series [276] and others. Because of low resistance of azo-group to reducing medium of polymer melt the majority of azo-dyes fail. However, it should be noted, that, in spite of the variety of synthesized dyes, their use for direct introduction into the polymer is limited owing to inorganic salts impurities. At light stabilization of polymers it is very important for dyes and additives being introduced to be more lightfast than the polymer, otherwise at the material service the process of deterioration of its physico-chemical characteristics in comparison with undyed one accelerates.

There are many works of patent character in which methods of obtaining PETP-fibres stock dyed [275-279] are described. Pigments (black and TiO_2) in the form of dispersions and pastes and additives soluble in polymer melt [280] are used for dyeing and stabilization.

In the work [33] it is shown that coloured synthetic polymers obtained by chemical modification directly during the process of their production by using aromatic compounds with different functional groups possess not only high colour strength but also high thermal stability.

Geller [281] used derivatives of oxyxanthane, containing six-member chelate cycle with intramolecular hydrogen bond for dyeing. Allen with his research workers used derivatives of amine antroquinone and oxyantroquinone as dyes. Krichevskiy [274] studied the effect of dispersed azo- and antroquinone dyes on physico-chemical properties of polyesters. In the work [282] direct turquoise lightfast and in the work [84] phthalocyan polyazodyes were used for PETP stock dyeing. Obtained fibres possessed high resistance to the effect of heat and light.

Increase of polyester fibre light stability is observed at presence of electron-acceptor para-substituents in the derivatives of 4 amine benzene.

Allen and his research workers found out that substituent location influences stabilizer efficiency. Cholles [283] indicated that substituent nature influences its efficiency.

Majority of dispersed dyes, used for PETP dyeing, are instable themselves [284, 286]. That is why it may be supposed that increase of dye concentration leads to accumulation of being formed radicals in the thin surface layer of the sample without mixing, as a result of which the rate of chain break rises and suppresses chain photooxidation of polymer. This effect was called by the authors [175] effect of concentration inhibition, which was observed in the case of polycaproamide light stabilization by action dyes.

Since there are data on stabilizing action of dyes at photo- and thermal destruction of PETP then it becomes very interesting to introduce such dyes-stabilizers by the method of stock dyeing, that is to introduce the dye into polymer at any stage of its production or fibre forming.

Polymer stock dyeing through the stage of obtaining dye polymeric concentrate (DPC) [168, 283] is in the centre of attention. Methods of dyeing at polymer synthesis and by the way of introducing dyed composition into polymer melt are combined in this method. DPC obtaining is carried out in the process of polymer synthesis in the presence of large amount of the dye (up to 50% of polymer mass).

Both initial granulate and ground production wastes of polymer may be used according to this method of dyeing. Dyeing may be fulfilled according to both periodic and continuous schemes. Hence, more uniform distribution of the dye in polymer mass is achieved and this gives the possibility of obtaining better monotone fibre dyeing during the whole technological process.

Since DPC is a composition of resin – dye with dyeing component content from 100 up to 50%, then the role of dye thermal stability is high, and it is desirable that resin nature in DPC should be the same as that of granulate being dyed.

For stock dyeing no more than 0,1-0,5 weight % is introduced, as such amount of dye cannot influence intermolecular interaction, chain regularity and polyester crystallinity [288].

At present new directions in modification are being developed using ultradispersed powders-fluorenes, being closed molecules, Cn, where n=60, 70, 80… They have crystalline structure and melt at the temperature above 237 ^{0}C, but still technology of their production is complex and inefficient [288]. Over the last years chemistry of polymers with conjugated system (PCS) developed into separate field of polymer science [289].

Problems of chemistry and physico-chemistry of PCS attracted attention not only of researchers working in the field of theoretical chemistry and physics but also researchers, working in the field of polymer modification [290].

Compounds, in whose molecule frequent repetition of groups and bonds, causing delocalization of valent electrons, is realized, are called polyconjugated systems.

A great number of substances with conjugated system, differing in type of conjugation and also the nature of heteroatoms and atomic groups in the main and side chains, has been synthesized at present. In this connection there has been suggested division of PCS into the following types [289]:

1 – PCS characterized by π, π – electron interaction in the main chains;
II – PCS characterized by interaction of π-electrons and non-distributed d-electrons;
III – PCS characterized by interaction of π-electrons with d-orbitals.

PCS of the II type are the most interesting as modifiers. Many types of polymers belong to this type. Their chains contain heteroatoms N, O, S with non-distributed π-electrons. Among them are: polybenzimidazoles, thiazoles, thiadiazoles, polymers of hydrocyanic acid; PCS with end heteroatoms – polymethynes, cyan dyes and their analogs, PCS with repeated heteroatoms in conjugated chains – polyazines, polyschiff bases, polymeric azo- and diazocompounds and so on.

The most widespread modifiers for polymers are salycyddienamines, which play the role of light stabilizers and deactivators of metals being in the polymer.

It is known that aromatic substances, containing heteroatoms of nitrogen, may have modifying effect on different polymers. Some derivatives of carbosol, according to literature and patent data, may be both light stabilizers and antioxidants in oxidative destruction of polymers.

From the series of highmolecular compounds the most interesting are polyconjugated azomethine compounds (PAC), stabilizing activity of which has been studied in the works [291, 292]. It is proved in these works that stabilizing effect is achieved at relatively small concentrations of azomethines. In the work [162] the author comes to a conclusion that at small concentrations azomethines work as inhibitors of radical processes, besides, they are ultra-violet absorbers. It should be noted that azomethines display light protective action both in vacuum and in the air.

Inhibitive action of PCS data is connected with the presence of both conjugated carbon-nitrogen bonds and end amine groups, moreover inhibitive activity of azomethines increase as conjugated chain grows longer [293].

For intensification of PCS lightstabilizing action there were obtained PCS with end amine groups, inhibiting by which, as it is supposed, proceeds according to the following scheme:

$R^* + R_1N_2 \rightarrow RH + R_1N^*H$;

where R – polymer radical.

Shielding action of azomethines is explained by the fact that they, owing to conjugation, turn luminous energy into heat energy or some other energy, which does not influence the process of polymer photochemical transformation. That is why, the longer is the conjugation chain the stronger is shielding ability of PCS.

Oligomer products of diamine condensation with carbonyl compounds, using excess of one of bifunctional compounds have been used lately. These products are used as stabilizers of thermooxidative destruction for different polymers, moreover oligomer stabilizers are more interesting because they are washed out from the polymer at photofading and sweated out, while in use least of all [294].

In the work [177] compounds on the basis of isomeric dianhydrides 4,4' bis – tetraphenyl (4,5 dicarboxynaph – 1-yl) phenyl benzophenone, and in the work [295] – sulfonamides were used as additives of polyfunctional action.

From the ecological and economic points of view it is necessary to search for such compounds which would combine dyeing, stabilizing, plasticizing and other properties, that is, it is necessary to search for additives of polyfunctional action.

Thus, information on modifying additives of polyfunctional action being used at present is represented mainly by patent works. Inorganic and organic compounds such as black, ultra-violet absorbers on the basis of benzophenone derivatives, inhibitors of radical processes-piperidines and stable nitroxyl radicals and so on – mainly colourless compounds, are recommended as modifying additives. However, problems of the effect of modifiers on photo- and thermal destruction of PETP are not completely interpreted in literature. Besides, different researchers come to different conclusions.

In some works direction of using different dyeing additives, including organic dyes, as PETP modifiers is being developed. Optical bleaches, disperse dyes, dyes of anthraquinone series, azo and phthalocyan dyes have found their application for this purpose.

At the same time stabilizing activity of dyes depends to a great extent on the method of their introduction into PETP composition.

Since the products from polyesters are obtained from the melt then suitable assortment of such dyes is highly limited and this defines the search of new compounds possessing high thermal stability and resistance to agressive reducing medium of PETP melt. Economic advantages of additives introduced into polymer mass compared with other dyeing methods should be also noted here.

PCS are the most interesting from the series of potential highmolecular modifiers. However, problems of such modifiers effect on polymer properties are not fully studied in literature.

From the economic and ecological points of view it is more reasonable to search for compounds with polyfunctional properties: thermal stabilizer, light stabilizer, antioxidant, plasticizer, dye and so on.

Literature survey has shown, that there are no data on the effect structure of introducticing modifiers on physico-chemical properties of polymer and also in what classes of chemical substances it is necessary to search for potential light-thermal stabilizers.

As it is shown earlier, effective additives-modifiers of polyfunctional action may be compounds, containing conjugated systems (CS) with high degree of conjugation. Hexaazocyclanes (HC), having developed conjugation chain, high photo and thermal stability, have been chosen as the objects of investigation. Presence of chromophore groups allows to use them in addition as polymer dyes.

Compounds, presented in Figure 3.1, were synthesized on the basis of phthalodimitrile (1); π-phenyldiamine (2) and derivatives of diaminefluorene (3). In order to check the effect of macrocyclic stabilizer structure on HC properties there were studied substances being the products of condensation of phthalodinitrile (1) with m-phenyldiamine (4).

Chemical structures of chosen hexaazocyclanes differ in degree of conjugation.

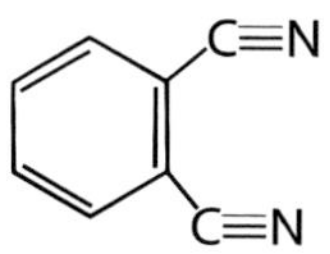

(1)

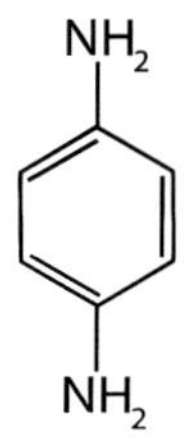

(2)

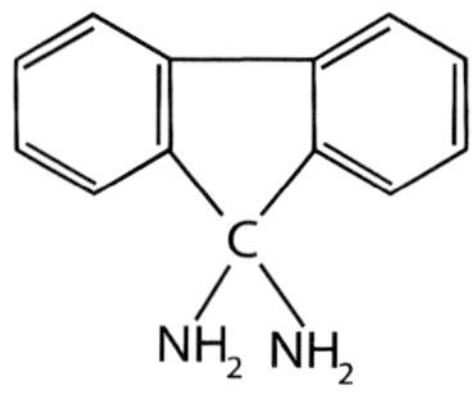

(3)

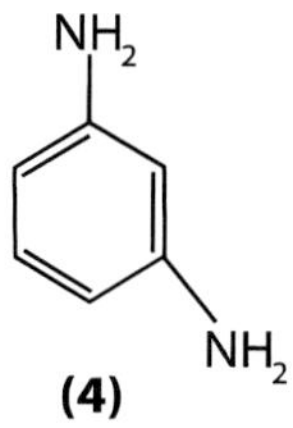

(4)

Characteristics of obtained compounds are given in Table 27.

Table 27. Characteristics of hexaazocyclanes

Conventional sign	Medium molecular mass	λ_{max}, nm (solvent acetone)	Melting T, ^{0}C	Decay T, ^{0}C	Colour of the dye
HC-1	600	360	305	330	yellow
HC-2	580	380	327	345	lemon
HC-3	670	340	315	327	yellow
HC-4	454	400	345	330	beige
HC-5	554	330	335	365	yellow
HC-6	552	360	327	350	beige
HC-7	538	330	300	347	light-brown

Electronic absorption spectra of compounds being used contain absorption band with λ_{max}=330-400 nm (Figure 3.2).

Criteria of compounds thermal stability are temperature values: melting, beginning of decay and losses of system initial mass at differential-thermal analysis (DTA) by 5, 10, 25%. These data are given in Table 28.

As it is seen from Table 28 all the compounds under investigation possess high thermal stability, exceeding greatly thermal stability of PETP. Temperatures of melting and beginning of decay of hexaazocyclanes are in the region of T_{melt}=300-345 ^{0}C and T_{dec}=327-365 ^{0}C; temperatures of check points of mass loss are also very high, that shows potential possibility of introducing compounds HC-1 – HC-7 into PETP melt, as their thermal stability exceeds temperature of PETP production and forming.

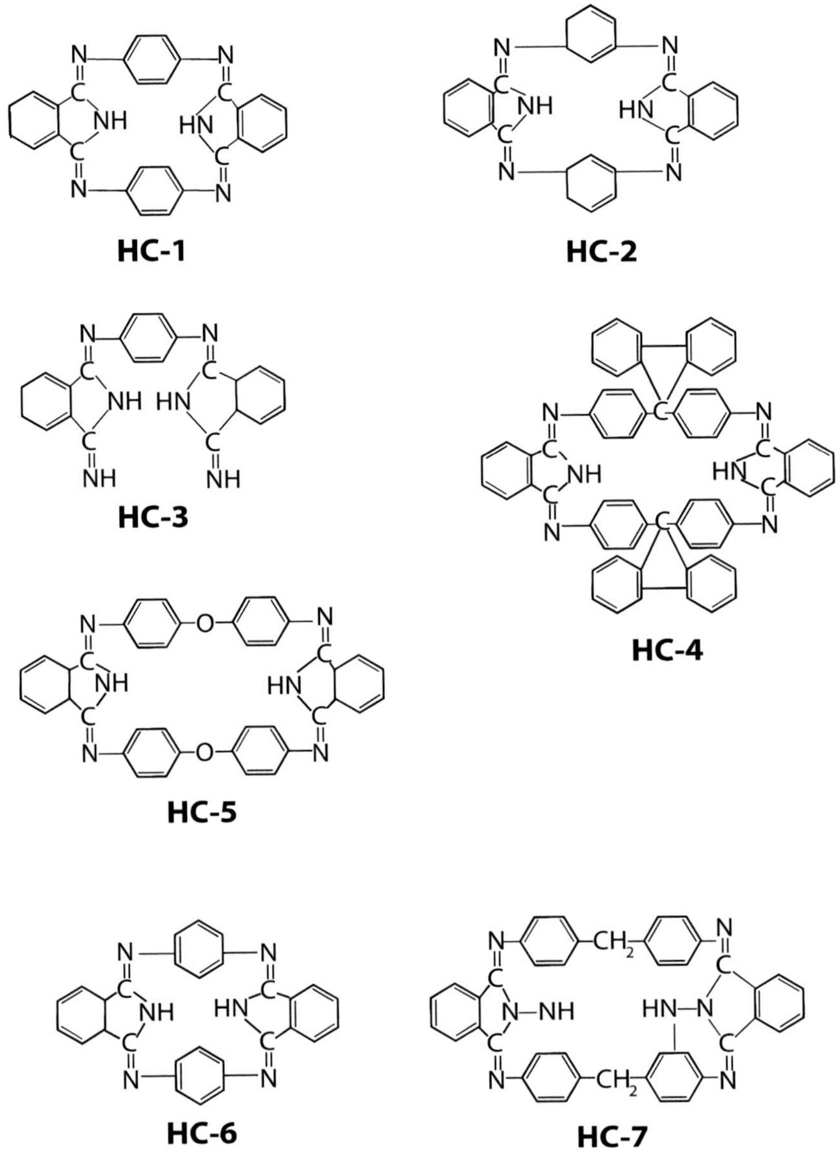

Figure 3.1. Structural formulae of hexaazocyclanes being used.

Table 28. Thermal characteristics of used hexaazocyclanes and initial polyethylene terephthalate

Compounds	Melting T, ^{0}C	T of beginning of decay, ^{0}C	T of mass loss, °C		
			5%	10%	29%
HC-1	305	330	435	480	552
HC-2	327	345	437	467	518
HC-3	315	327	409	455	507
HC-4	345	330	415	449	480
HC-5	335	365	469	490	539
HC-6	327	350	440	500	553
HC-7	300	347	425	486	586
PETP	269	320	360	415	428

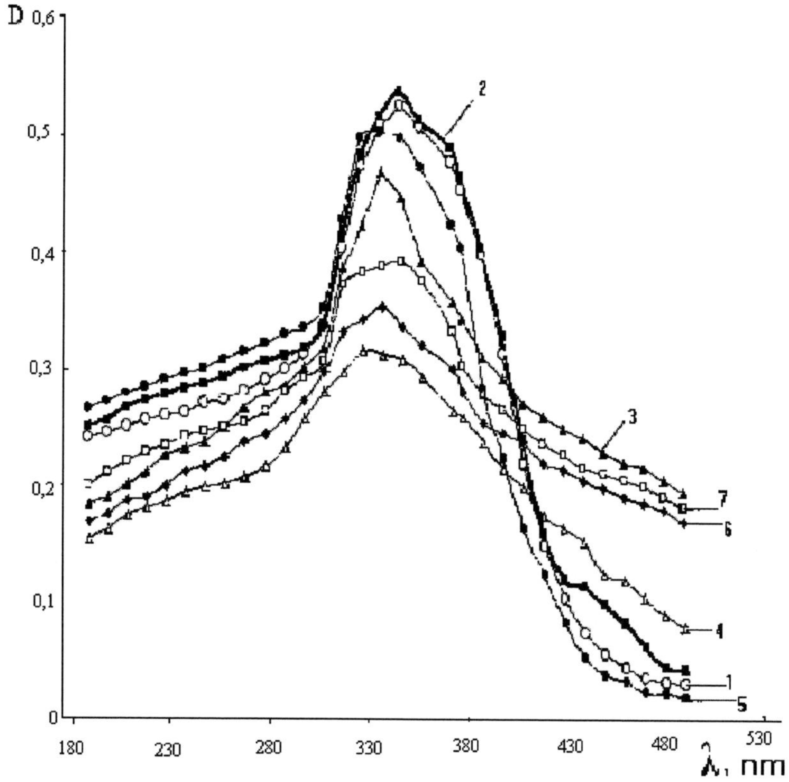

Figure 3.2. Electronic absorption spectra of hexaazocyclanes HC-1 (1), HC – 2 (2), HC – 3 (3), HC – 4 (4), HC – 5 (5), HC – 6 (6), HC – 7 (7).

3.4. Modification of PETP by Hexaazoncyclanes

One of the methods of introducing plasticizers, dyes and other additives into PETP is modification at the stage of synthesis.

Synthesis of PETP was carried out with additives HC-1-HC-7 in the amount of 1 mass %. Additives were introduced in the form of suspension in ethyleneglycol at the beginning of the process of peretherification.

Synthesized polymers were uniformly dyed into proper colours, intrinsic viscosity was in the region 0,66-0,77 while the same index for colourless PETP, obtained under the same conditions, was 0,62, that also corresponds to technological parameters.

The process of fibres forming, containing additives HC-1 – HC-7 was stable and there were no difficulties at drafting. Being obtained fibres were drawing easily, they had the same thickness along the whole length and there were no any swellings and non-uniformities. Fibre breaks during formation appeared seldom. Here one may come to a conclusion that introduction of hexaazocyclanes did not influence the process of formation and so introduction of these compounds into reaction mass does not practically influence fibre-forming properties of PETP.

Miscibility of additives being used with polymer and uniformity of obtained polymer were controlled by comparison of spectra of hexaazocyclanes solutions absorption in cresol and solutions of dyed fibres before and after spinneret (Figure 3.3). Spectra practically coincide with each other, showing full solution of dye and good homogenization of polymer melt.

Hexaazocyclanes, depending on concentration, dye polymer from yellow up to brown colour. It is characteristic that at the concentration of the additive 1% of the mass polymer becomes bright coloured. Correlation of conditions for obtaining initial and dyed PETP fibres has shown that introduction of HC-1 – HC-7 into PETP melt in the amounts from 0,5 to 1% does not decline the rate of fibre forming, does not cause their adhesiveness, fragility and early solidification. This is undoubted advantage of suggested additives as many substances, being potentially dyes, cannot be used as they are not compatible with technological processes of producing PETP-fibres.

In order to understand special features of additives effect on microcrystalline polymer structure it is necessary to pay attention to features of the structure of PETP – polymer blocks.

Polymer, as solid, has imperfect supermolecular structure; it has amorphous regions, regions without structures and crystalline regions with well packed macromolecules. Lamellar monocrystals (lamellae) in which macromolecules are placed perpendicular to wide surface of a plate are the main structure of polymer crystalline part. Usually length of macromolecule being crystallized greatly exceeds thickness of lamella and, to be placed into the crystal, macromolecule must repeatedly fold.

PETP crystallizes easily, moreover in the work [129] it has been shown that PETP crystallization takes place at the expense of transformation of groups –O-CH2-CH2-O into transconfiguration:

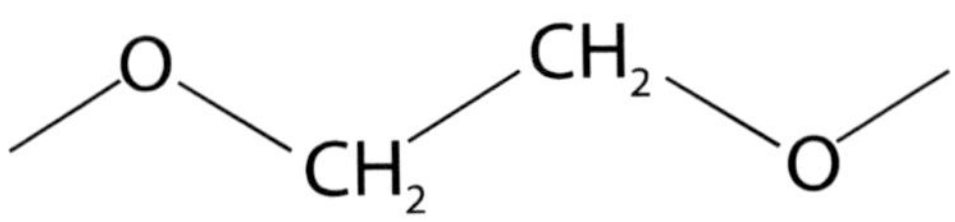

Probably it is easier for PETP molecule to fold in order to be packed into crystal. Since PETP is produced from the melt then packing of macromolecules into crystal occurs irregularly, as packing of macromolecules and growth of crystals are controlled by local mobility of macromolecules elements near growing surface of the crystal [295].

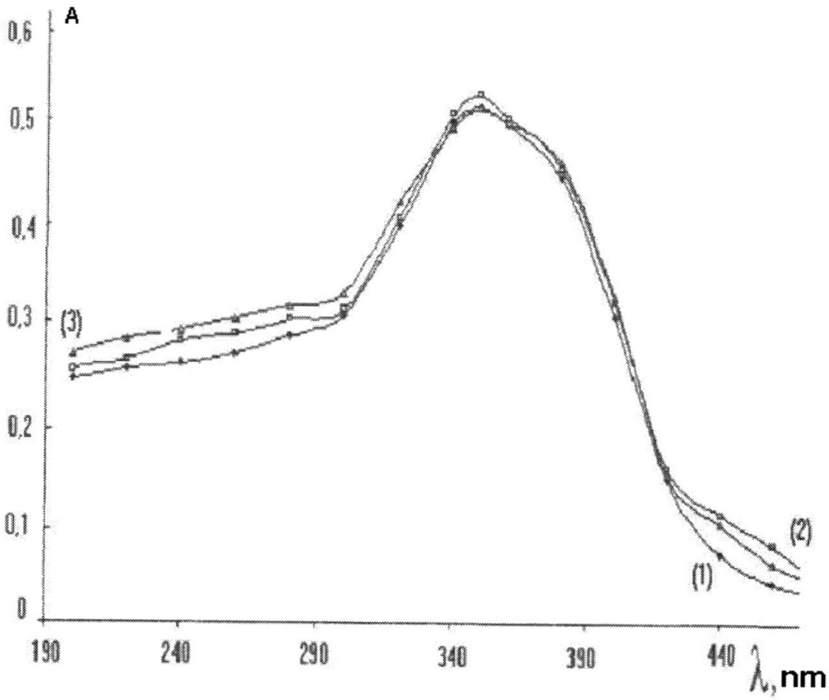

Figure 3.3. Spectra of absorption of dye solutions HC-1 (1) and PETP dyed by this dye before (2) and after (3) spinneret in cresol.

Space between microcrystals is filled by uncrystallizable part of polymer substance, which grouts microcrystals combining them into fibrills. Fibrills are fibered structures beginning in the centre of the crystal and being oriented radially. These are fibrils which organize lamellae and provide their radial orientation.

Monocrystals (lamellae) and fibrils, in their turn, are being organized into larger structural elements – spherulites. Lamellae are organized in spherulite in such a way that they line up along radial directions from the centre of spherulite.

Amorphous region is formed from polymer parts, which do not crystallize because of unfavourable location of segments.

If crystalline sectors contain PETP macromolecules in which groups –O-CH2-CH2-O- are in transconfiguration, then in amorphous part this group is both in trans- and cysconfiguration.

Amorphous phase of polymer is concentrated between crystals, fibrils and spherulites. It consists of loops of different length, adhesions, passing chains of different length and intensity, chains ends, uncrystallizable macromolecules or their sectors, containing branches, stereochemical defects and irregularities in molecular structure. Lowmolecular fractions, impurities of lowmolecular substances and so on are displaced into amoephous part of polymer during the motion of crystallization frout. During copolymers crystallization amorphous part is composed from sequences of component monomer links, incapable to regular laying and crystallization.

Deformation of polymer material, which, in the first place, affects amorphous regions, takes place during its mechanical loading. First, there occur conformational transformations of passing molecules, and then, when reserve of their conformations is being partly exhausted, there begins mechanical deformation of valence angles and chemical bonds with their further break.

Thus, amorphous regions of polymer, bounded by crystals, are "macroreactors" in which chain process of mechanodestruction and failure of material is being developed. Hence, decrease of the amount of polymer amorphous part will lead to strengthening of the fibre and improving its physico-mechanical characteristics.

Method of microphotography was used to study polymer structure and effect of additives HC-1 – HC-7 introduction. Microphotography of polymer fibres shear was carried out with the help of microscope "Polar", at enlargement by 40 times, the microscope had special extension for photography. Photoes of dyed and undyed fibres and also photos of their longitudinal shears are given in Figure 3.4. – 3.10.

From the analysis of microphotoes it follows that introduction of hexaazocyclanes does not disturb uniformity of polymer structure that shows good solubility of hexaazocyclanes in PETP melt. On the basis of the technique [167, 296] according to microphotoes there has been performed calculation of heterogencous embedding of additives into fibre which gives the value 125 units per kilogram, which is lower than 300 units per kg permissible by State standard. This shows the uniformity of additive distribution in polymer.

Microphotoes of initial PETP – fibre and its cross-section are given in Figure 3.4. Two regions – amorphous and crystalline – are observed in the photo of cross – section of initial PETP fibre (Figure 3.4 b). Lamellar and fibrillar formations are seen in crystalline regions, but they are not quite developed. Amorphous part occupies large volume of the general polymer space.

Microphotoes of PETP – fibres, modified by HC-1 – HC-6 and their cross-sections are shown in Figure 3.5. – 3.10. Comparing microphotoes of initial and modified PETP it should be noted that modified PETP displays large tendency to crystallization.

Lamellar and fibrillar formations are seen in the photos of fibres dyed by HC-1, HC-3, HC-4 (Figure3.5, 7, 8), moreover these formations are large in dyed fibres than in initial PETP.

Tendency to crystallization in fibres with additives HC-2, HC-5, HC-6 increases still more. Radial spherulite formations of rather large size are observed in microphotoes of fibres modified by these hexaazocyclanes (Figure 3.6, 9, 10).

Two processes of formation of new phase nucleus and further growth of already present ones take place at the moment of crystallization. Nuclei formation is the first act of nucleation.

In molten state PETP consists of associates which may be the cores of formation of crystallization nuclei. Completion of crystallization process and crystal size depend very much on rates chains lay-up on the side of growing crystal and relaxation of large segments relation.

Increase of modified PETP ability to crystallization compared with unmodified PETP, as we see it, is connected with the fact that hexaazocyclanes play the role of singular centres of crystallization. This causes change of molecular orientation in the fibre at the moment of formation, when supermolecular structure begins to organize. And here new more regulated structure is being formed and this agrees with conclusions of the works [130,297].

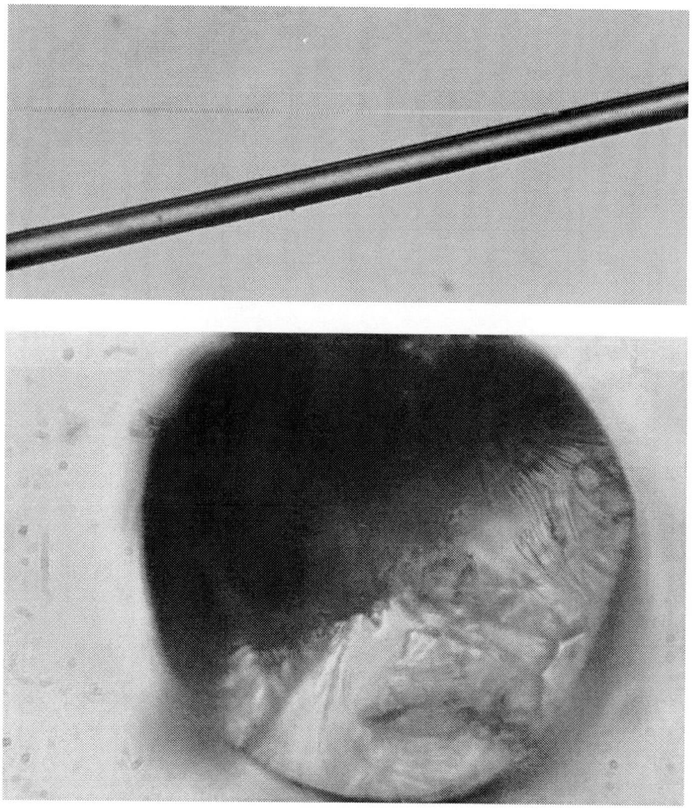

Figure 3.4. Microphotographs of PETP – fibre (a) and its cross-section (b) without introduction of hexaazocyclanes.

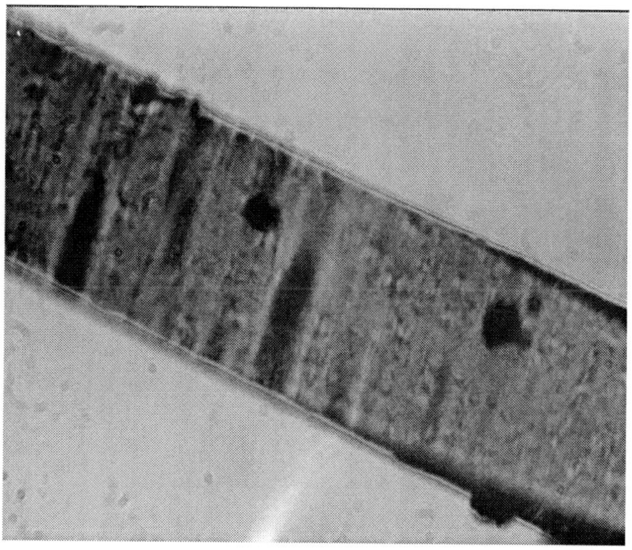

Figure 3.5. (Continued on next page).

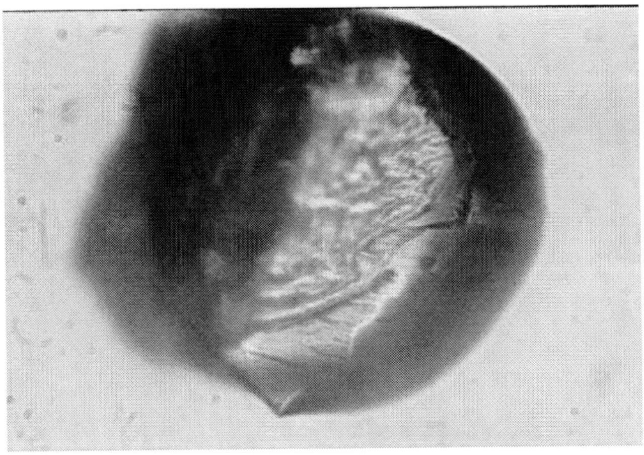

Figure 3.5. Microphotographs of PETP – fibre (a) and its cross section (b), modified by HC-1.

Figure 3.6. Microphotographs of PETP – fibre (a) and its cross-section (b), modified by HC-2.

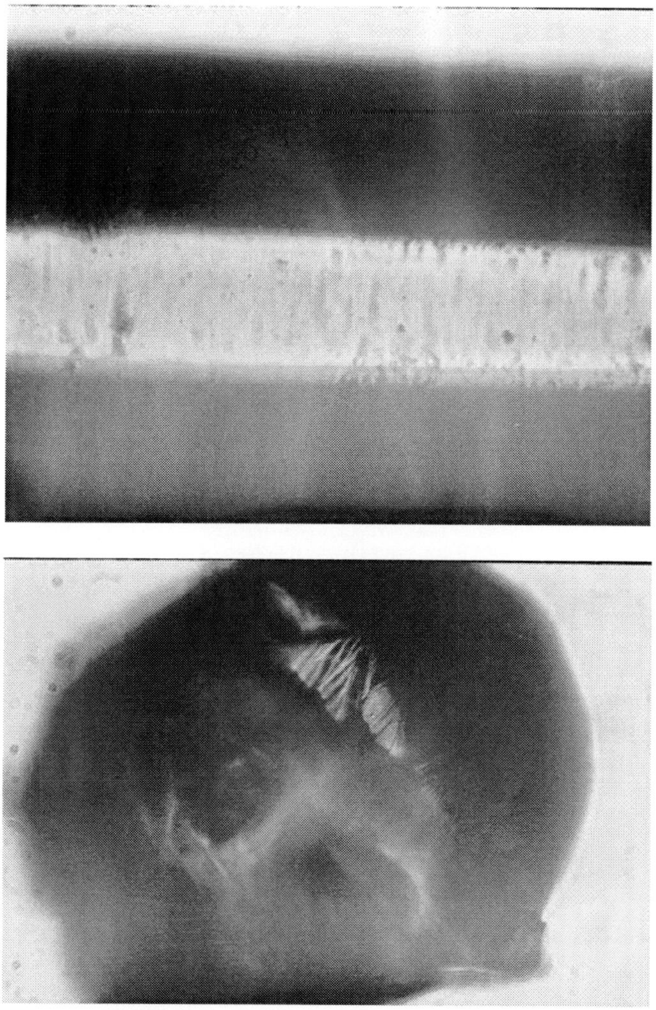

Figure 3.7. Microphotographs of PETP – fibre (a) and its cross-section (b), modified by HC-3.

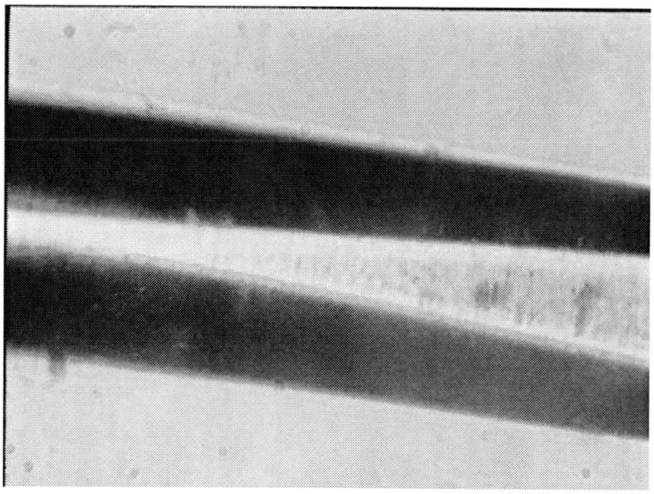

Figure 3.8. (Continued on next page.)

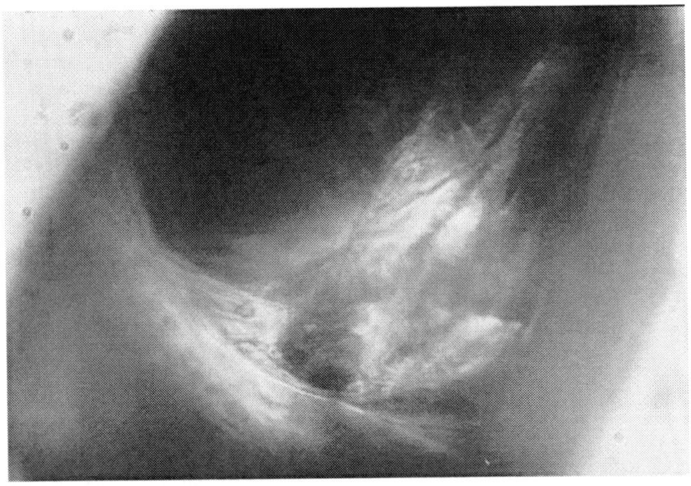

Figure 3.8. Microphotographs of PETP – fibre (a) and its cross-section (b), modified by HC-4.

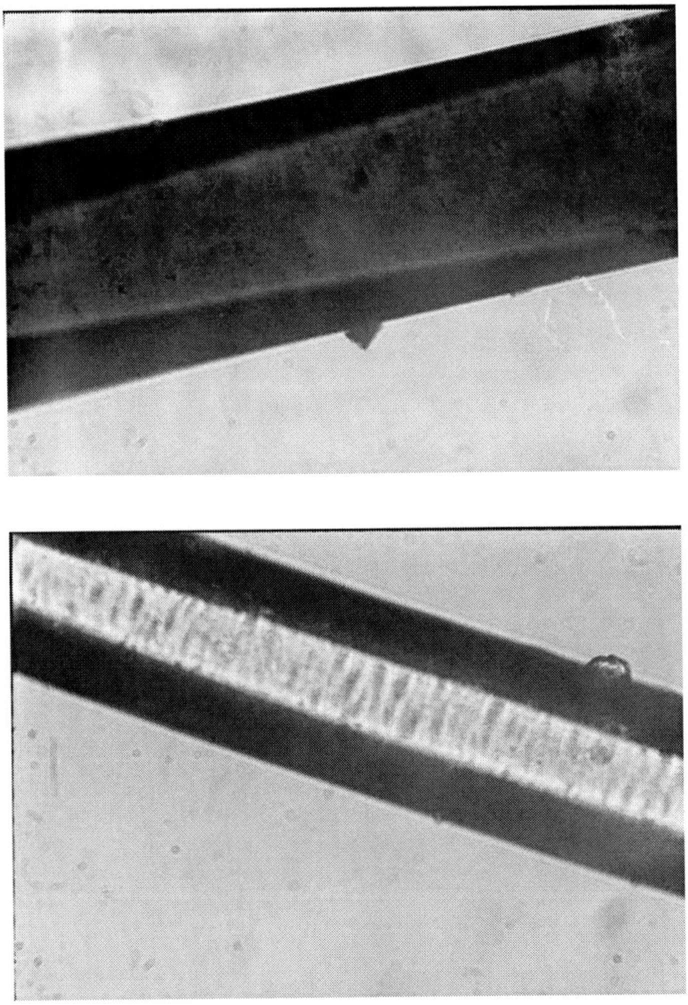

Figure 3.9. Microphotographs of PETP – fibre (a) and its cross-section (b), modified by HC-5.

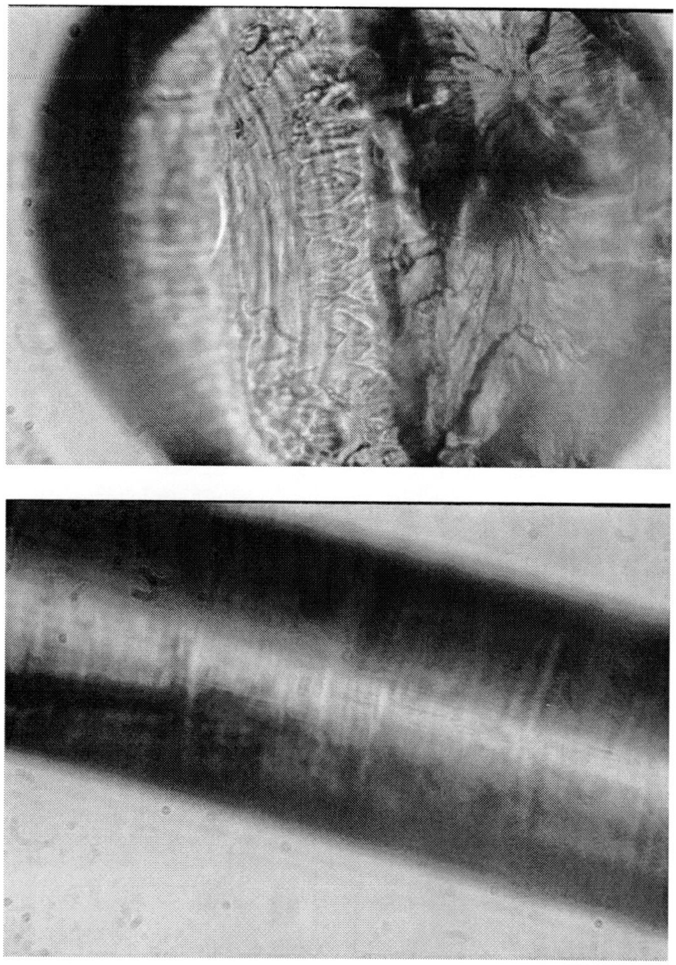

Figure 3.10. Microphotographs of PETP – fibre (a) and its cross-section (b), modified by HC-6.

Authors of these works consider that hexaazocyclanes influence the rate of formation of crystallization nuclei at the expense of surface energy of crystallites that leads to greater regularity of structure and decrease of critical sizes of crystallites.

Comparison of microphotoes of modified and unmodified PETP – fibres shows that introduction of hexaazocyclones increases the degree of PETP crystallinity that well agrees with above-mentioned literature data.

Amount of amorphous part in modified fibre decreases; this will increase fibre strength and improve its mechanical characteristics, since amorphous region is the most weak for mechanical loads action and stress as it contains regions of the least degree of order in macromolecules package and in the region of defects concentration.

High degree of crystallinity causes high fibre density, resistance to the action of chemical reagents and physico-chemical properties of polymer. Fibrillar structure of the fibre is seen in microphotoes of modified fibres, moreover introduction of hexaazocyclanes into PETP facilitates increase of regularity and density of polymer structure. All this proves that introduction of hexaazocyclanes additives increases PETP ability to crystallization.

Obtained samples of PETP – fibres being investigated were completely solved in solvents without formation of precipitates (possible products of hexaazocyclanes and PETP interaction), that shows attaining required inertness of components regarding each other according to technology.

Thus, all compounds being tested are solved and combined with PETP melt quite well, stable in polymer melt, do not have negative effect on polymer, facilitates polymer crystallization and fibre strengthening and do not require change of factory technological regulations.

3.5. Thermo- and Thermooxidative Destruction of Modified PETP – Fibre

Data on complex thermogravimetric analysis at aerodynamic heating are given in Table29.

Table 29. Thermal characteristics of PETP – fibres dyed by hexaazocyclenes

Fibre sample	T of melting, °C	T of initial oxidation, °C	T of mass loss, °C		
			5%	10%	25%
PETP - initial	269	320	360	415	428
PETP + additive HC-1	250	335	441	444	447
PETP + additive HC-2	245	385	480	526	576
PETP + additive HC-3	260	265	345	360	370
PETP + additive HC-4	250	290	360	368	375
PETP + additive HC-5	260	340	400	480	505
PETP + additive HC-6	263	340	420	445	453
PETP + additive HC-7	250	365	415	440	503

DTA curves of modified and unmodified PETP – fibres taken in air at the rate of temperature rise 5 °C/min are given in Figure 3.11 – 3.13.

Curve of differential – thermal analysis (DTA) of unmodified PETP on which there given physical transitions and chemical transformations, accompanied by exothermanl and endothermic effects, which correspond to separate bend of a curve is shown in Figure 3.11 (curve 1).

Exothermic processes are crystallization, oxidation and decay (sections В, Г, Д in Figure 3.11), while vitrification and melting are endothermic transformations (sections А, Б in Figure 3.11).

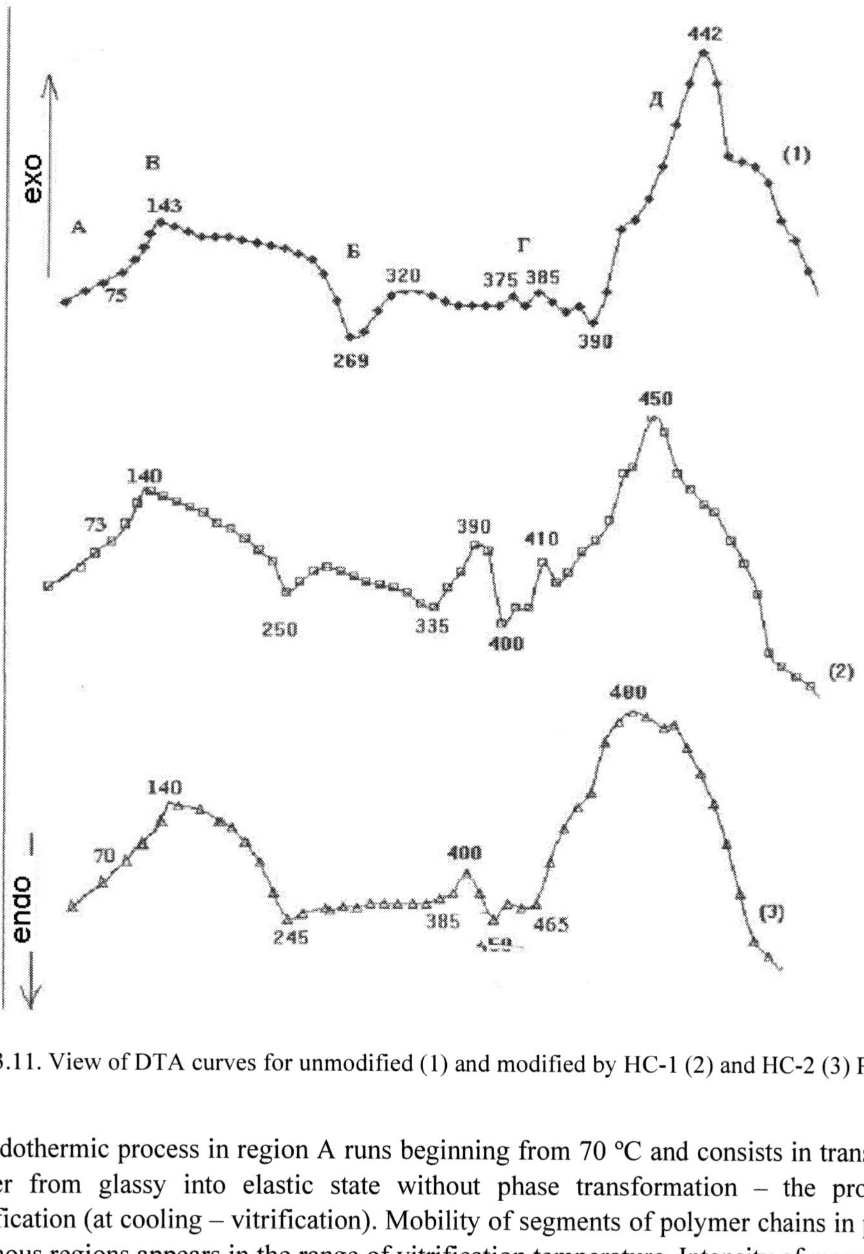

Figure 3.11. View of DTA curves for unmodified (1) and modified by HC-1 (2) and HC-2 (3) PETP – fibres.

Endothermic process in region A runs beginning from 70 °C and consists in transition of polymer from glassy into elastic state without phase transformation – the process of devitrification (at cooling – vitrification). Mobility of segments of polymer chains in polymer amorphous regions appears in the range of vitrification temperature. Intensity of movement in polymer amorphous regions decreases with the increase of crystallinity. Mobility of molecule parts, situated in amorphous regions, is limited at the expense of the fact that other parts are a part of crystalline region. Another reason for decrease of molecules mobility in amorphous phase is probably stress which the fibre is probably subjected to.

Exothermic process of phase transition – crystallization, characterized by peak in the region Б at the temperature 143 °C runs after PETP devitrification.

The second endothermic process (region B) is observed at the temperature higher 230 °C and ends at 270 °C. And during this process crystalline structure breaks and polyester melts. This transition, as crystallization, is phase. High temperature of PETP melting is described by specific role of aromatic nucleus linked by p – para-position [245].

On the thermogram of unmodified PETP, Figure 3.11 (curve 1) temperature range of melting is not sharp transition by itself, but the region of extended range. This points to the presence, together with absolute order, of metastable crystalline forms extending temperature range of melting. Initial PETP and modified by hexaazocyclones have clearly defined peaks of melting (endoeffect at 250-269 °C).

Clear character of melting peaks points to narrow distribution of polymer crystalline part according to crystals sizes [298]. However, endoeffect at 269 °C (initial PETP – fibre), not accompanied by mass loss and corresponding to the melting of crystalline regions, in modified PETP – fibres, containing additives of hexaazocyclanes HC-1 – HC-7, is shifted into the region of lower temperatures (250-263 °C).

According to Bell and Damblton [299] PETP may have two morphological forms of crystals which define double endothermic effect in melting region. Folded structure was assigned to form I, which initial more high temperature endotherm corresponds to, and for the form II, responsible for more low temperature endotherm, was suggested crystalline structure from more extended chains.

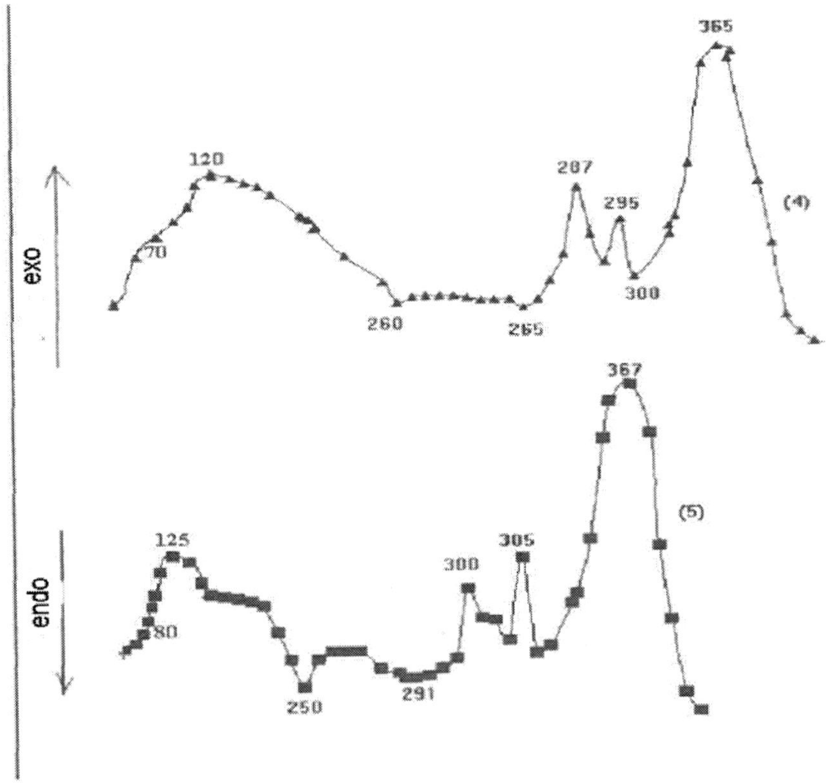

Figure 3.12. View of DTA curves for PETP – fibres modified by HC-3 (4) and HC-4 (5).

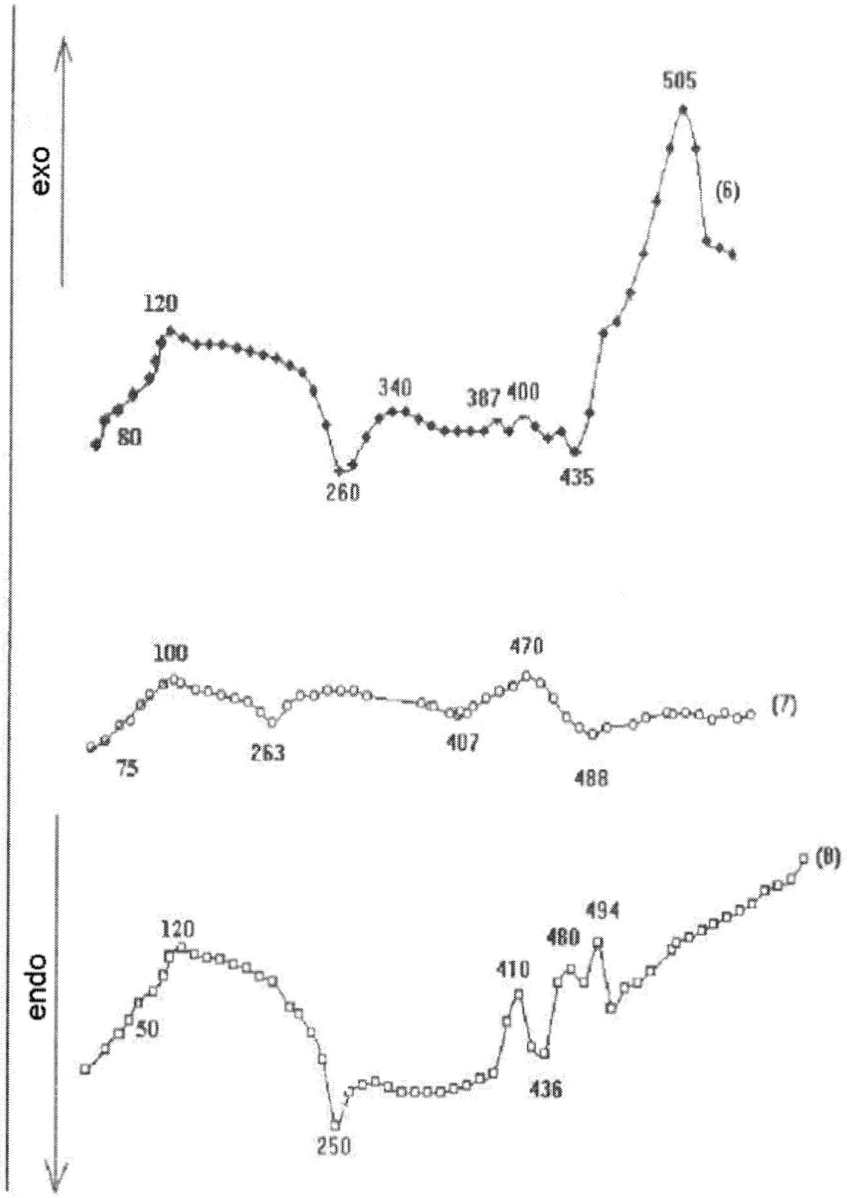

Figure 3.13. View of DTA curves for PETP – fibres modified by HC-5 (6), HC-6 (7) and HC-7 (8).

Nilomn [300] has supposed that the length of form I folds causes the degree of perfection of crystalline form II being formed by the way of partial unfolding of form I folds, though their nature has not been cleared yet.

Analysis of microphotographs of obtained PETP – fibres samples has shown that there are both morphological forms of crystals in obtained polymer.Prevalence of one of morphological forms will probably define the value of melting temperature [300].

On DTA curves of initial PETP – fibre and modified by hexaazocyclanes (Figure 3.11, curves 2,3; Figure 3.12, 3.13), melting peak is the region having one maximum that proves made earlier supposition about dependence of melting temperature on morphological structure of polymer. Besides, comparison of DTA of initial and modified PETP – fibres points to the shift of the maximum of melting peak into the region of lower temperatures.

Hence, in modified fibres crystals have mainly morphological form II (extended polymer chains, combined into joints of crystallites), while in initial PETP crystals mainly have morphological form I (folded structure). That is why endothermic effect at the temperature 269 °C is observed for initial PETP, explained by melting of plane folded crystallites (morphological form I). In modified PETP – fibre this effect is observed at temperature 245-263 °C, it is explained by melting crystallite joints (morphological form II).

Intensity of endothermic peak, corresponding to melting region is qualitative criterion of polymer thermal stability [300]. In our case it is seen that intensity of melting peak of modified and unmodified polymer are the same. Hence, introduction of hexaazocyclanes at any rate, does not deteriorate thermal stability of PETP – fibres. Only thermal effects, connected with crystallite melting, are registered in thermograms during heating of PETP samples. On DTA curves of modified PETP – fibres it is seen that the range of melting peak in them is larger than in initial PETP.

Proceeding from above-mentioned one may assume that PETP – fibres modified by hexaazocyclanes possess greater degree of crystallinity. This proves the conclusion made earlier that hexaazocyclanes molecules being introduced become additional centers of crystallization, thereby increasing degree of crystallinity of modified PETP – fibre.

In addition to this, hexaazocyclanes additives cause decrease of melting temperature (Figure 3.11 – 3.13), that is hexaazocyclanes have plasticizing effect on PETP.

Kinetics of radical chain process of polymer thermal destruction includes stages of initiation, growth of reaction chain, chain transmission, its break. Reaction of chain transmission occurs mainly at the expence of hydrogen break from polymer chain.

PETP is rather stable at relatively low temperature regarding oxygen. However, oxidation process runs at considerable rate at the temperatures above 220-250 °C.

Oxidation of polymer is accompanied by the change of their structure – physical properties – crystallinity, molecular mobility, strength and so on. Orientation of crystalline regions is disturbed in stressed samples; the number of crystals with definite space orientation of crystal lattice axes decreases, the form of the curve of crystal distribution axes in respect to orientation axis changes. Change for the worse of crystals orientation is explained by stress relief, occurring in polymer at the expense of oxidative destruction of macromolecules in amorphous intercrystalline region [301].

Only a few works are devoted to the investigation of the mechanism of thermooxidative destruction process. This is explained by the fact that PETP visible oxidation begins at relatively high temperature (above 250 °C), when together with oxidative processes there are processes of purely thermal destruction. In the presence of oxygen it should be expected that the process of destruction will run according to radical – chain mechanism with initiating along the bond C-H at methylene group:

$$\mathbf{RH} \xrightarrow{O_2} \mathbf{R\bullet + HOO\bullet} \xrightarrow{O_2} \mathbf{ROO\bullet} \xrightarrow{RH} \mathbf{ROOH}$$

Oxidation of solid polymers is radical – chain process with marked branching and square break of kinetic chains, the main branching product being hydroperoxide. Fast – decomposing hydroperoxide is localized in amorphous phase, being more stable in crystals [302].

Hydroperoxide is not only the main branching agent but also forerunner of all lowmolecular products and breaks of molecular chains, leading to the change of molecular mass and molecular – mass distribution (MMD).

Oxidative processes are localized in amorphous interlayers, in interfibrillar regions and others. Crystallinity and crystals sizes increase at initial stages of oxidation [303]; it also means that oxidation is localized in amorphous part. Destructive decay of passing macromolecules in amorphous interlayers release them and facilitates folding of chains into crystals. Destruction and amorphicity of crystals takes place only at deep stages of oxidation. Solubility of oxygen in polymer depends not only on polymer crystallinity but on microstructure of amorphous or defect sections.

At samples heating in the presence of oxygen of the air exothermic and endothermic effects, corresponding to oxidation reactions and polymer destruction, are seen on DTA curves. In our case, much more differences than in melting region are observed on DTA curves in the region of 400-500 °C where complex process of oxidation and PETP decay take place.

As it has been found, microphotographs show that hexaazocyclanes decrease the amount of amorphous phase, that is why it should be expected that introduction of additives of these compounds into PETP ought to break polymer oxidation, as fastdecomposing hedroperoxide is in amorphous part of polymer.

Both decrease and increase of intensity and position of exothermic effect maximum at the temperatures above melting temperature, caused by intensive polymer oxidation are marked on DTA curves of modified PETP samples comparing with DTA curves of initial ones.

Increase of temperature maxima of oxidation peaks (Figure 3.11 curves 2, 3; Figure3.13 curves 6,7,8) is observed for PETP – fibres dyed by HC-1, HC-2, HC-5, HC-6, HC-7. Opposite effect – temperature of oxidation maximum decreases – is observed on DTA curves in the fibres with additives HC-3 and HC-4 (Figure 3.12 curves 4, 5). Besides, increase of intensity of decay peak and shifting of decay temperature in the direction of higher temperatures takes place in fibres modified by HC-1, HC-2, HC-5 (Figure 3.11 curves 2 and 3; Figure 3.13 curve 6). Increase of intensity of oxidation peak takes place on thermograms of the fibres, modified by HC-3 and HC-4 (Figure3.12 curves 4, 5), besides temperature of oxidation maximum decreases. Decrease of intensity of decay peak takes place in samples with additives HC-6 and HC-7 on DTA curve (Figure 3.13 curves 7, 8), though decay temperature increases in comparison with initial sample. Opposite effect is observed in fibres with additives HC-3 and HC-4 on DTA curves (Figure 3.12 curves 4 and 5) in comparison with fibres dyed by HC-6 and HC-7. Increase of decay peak is observed in fibres modified by HC-3 and HC-4, but in addition maximum of this effect is shifted in the direction of lower temperatures.

From the analysis of DTA curves one may come to a conclusion that introduction of hexaazocyclanes have different effect on thermal and thermooxidative stability of PETP. Supermolecular structure of fibres is little studied by itself and it is difficult to predict and explain its dependence on the structure of modifier at the given stage of investigation.

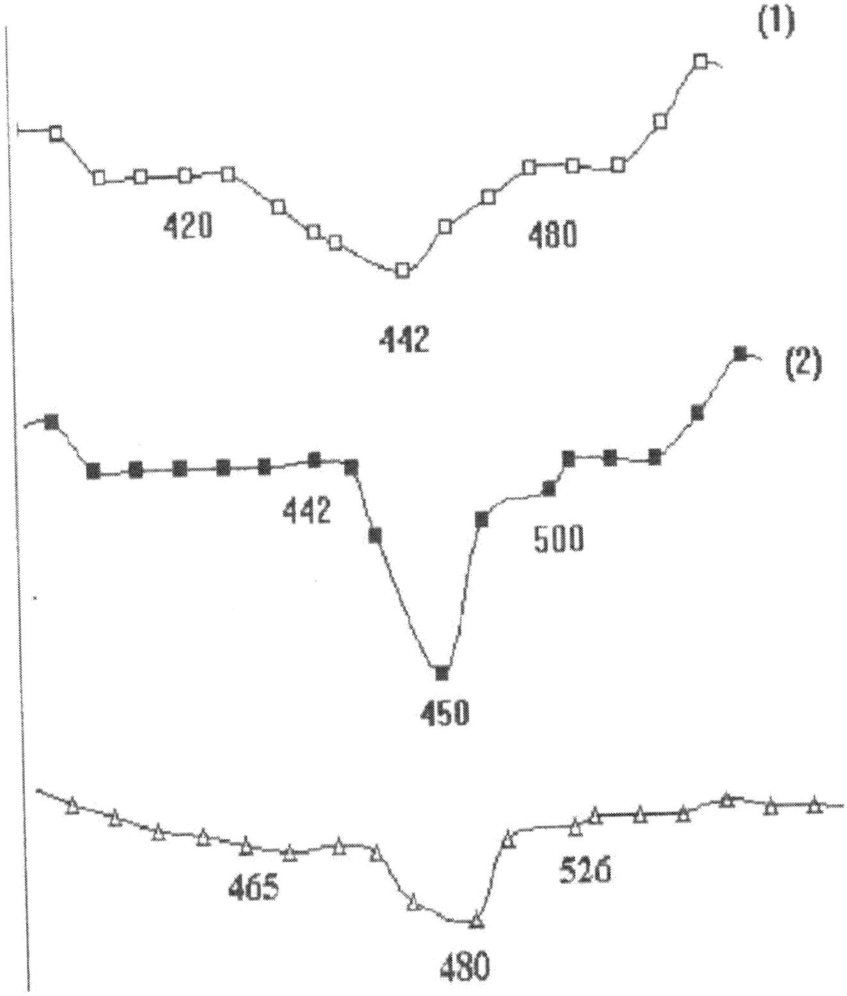

Figure 3.14. View of DTG curves of unmodified (1) and modified by HC-1 (2), HC-2 (3) PETP – fibres.

It was shown earlier that effective stabilizers of thermal and thermooxidative destruction should have conjugation system with developed chain of conjugation. Hexaazocyclanes being used possessed different degree of conjugation. Owing to this, all hexaazocyclanes additives being introduced had different stabilizing effect on polymer. This supposition is proved by differential thermal analysis of PETP – fibres modified by hexaazocyclanes. HC-3 (see Figure 3.12 curve 4) and HC-4 (see Figure 3.12 curve 5) effect on PETP differs from the effects of HC-1, HC-2, HC-5, HC-6 and HC-7. Probably it is connected with plane structure of additives.

Thermooxidative destruction of PETP is accompanied by libration of volatile products, so there is weight loss in polymer sample being tested. That is why, that temperature, at which decrease of polymer weight is observed, characterized its thermooxidative stability.

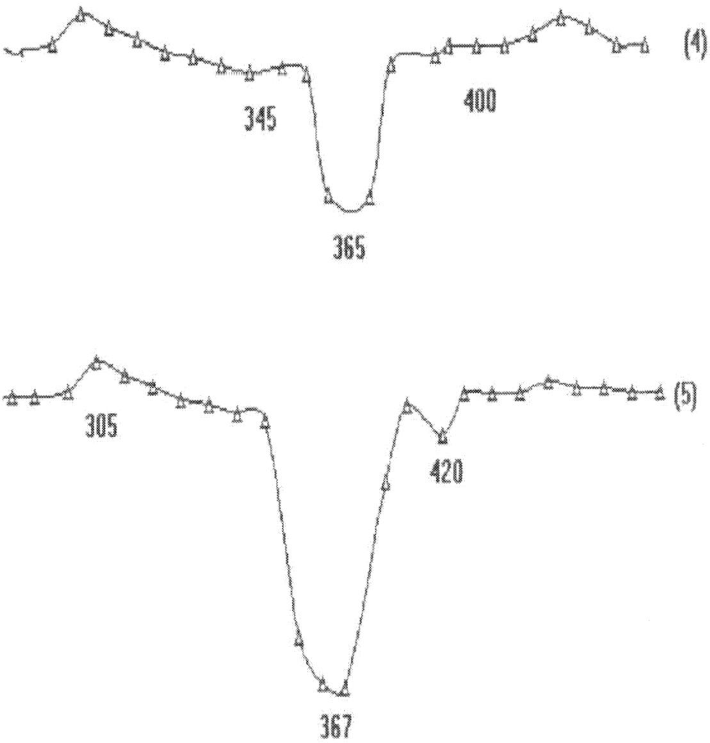

Figure 3.15. View of DTG curves of PETP – fibres, modified by HC-3 (4) and HC-4 (5).

Curves of differential thermogravimentric analysis (DTG) (Figure 3.14 – 3.16) were taken for definition of this temperature. Maximum rate of mass loss of initial PETP – fibre is observed at 442°C (Figure 3.14). In fibres modified by HC-1, HC-2, HC-5, HC-6, HC-7 this index shifts into the region of much higher temperatures (Figure 3.14 curves 2, 3; Figure 3.16 curves 7, 8). During introduction of additives into PETP – fibre (additives – HC-3 and HC-4) maximum rate of mass loss shifts in the direction of the low temperatures region.

Data of thermal stability of modified and unmodified PETP – fibres, obtained on the basis of TG, are given in Figure 3.17. These curves show, that thermal stability of PETP – fibres with dye – modifier introduction increases, as intensive decay of modified samples begins at much higher temperatures and decay depth at one and the same temperature decreases. HC-3 and HC-4 are the exception again. The greatest thermostabilizing activity is displayed by PETP – fibres modified by HC-2.

From the data of Table 3 it is seen that introduction of hexaazocyclanes into PETP increases temperature of the beginning of PETP decay by 15-70°C, HC-3 and HC-4 being the exception. Such effect is probably connected with structural changes in PETP, taking place at addition of HC-1, HC-2, HC-5, HC-6, HC-7 compounds into polymer, and possible inhibiting effect on thermooxidative destruction of PETP.

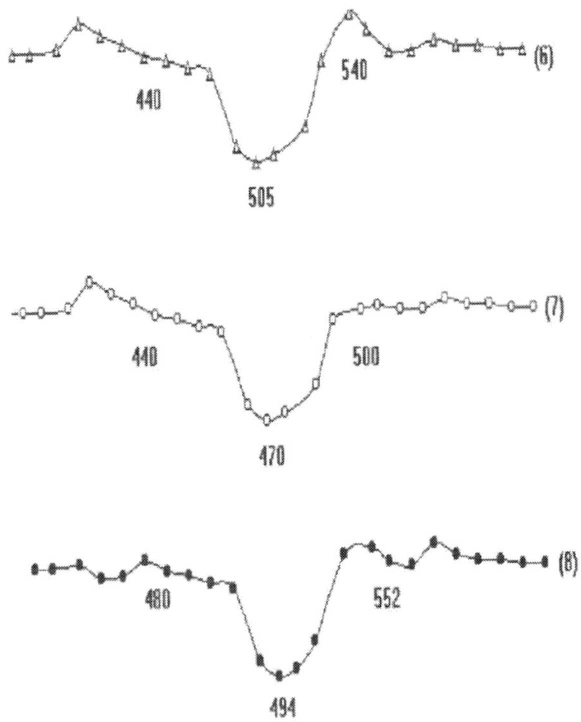

Figure 3.16. View of DTG curves of PETP – fibres, modified by HC-5 (6) and HC-6 (7), HC-7(8).

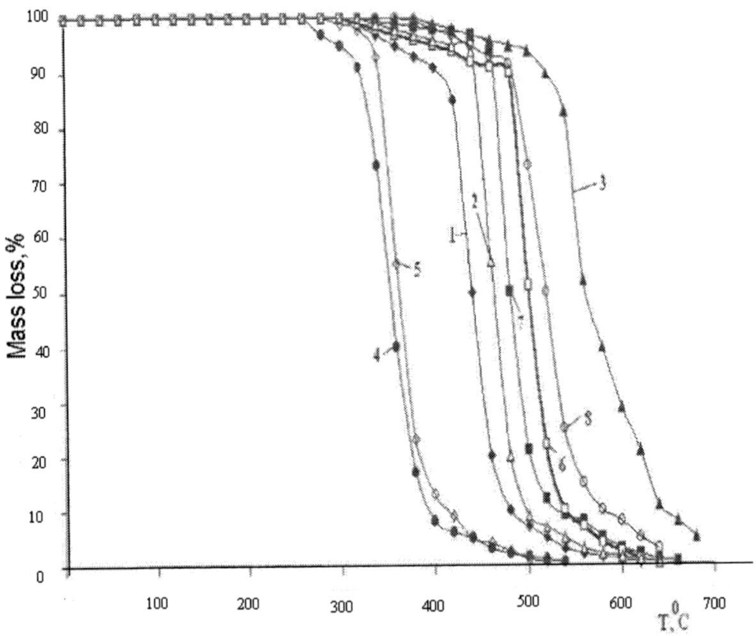

Figure 3.17. Thermal stability of unmodified (1) and modified by HC-1 (2), HC-2 (3), HC-3 (4), HC-4 (5), HC-5 (6), HC-6 (7), HC-7 (8) PETP – fibre.

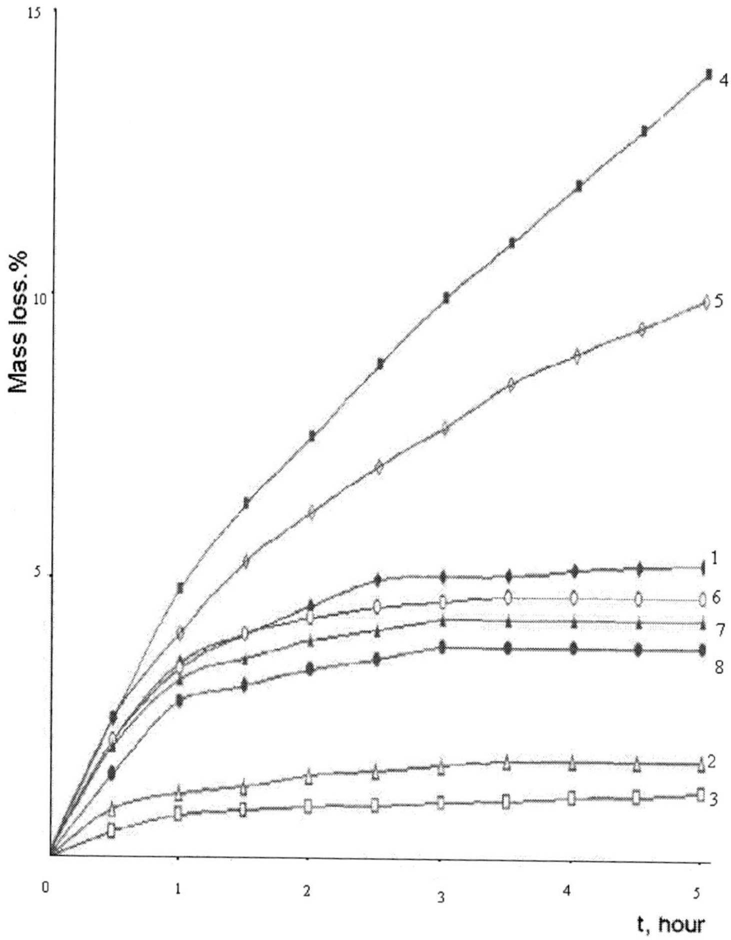

Figure 3.18. Kinetic curves of mass loss in isothermal conditions (240°C) of PETP – fibres: 1 – undyed; 2 – modified by HC-1; 3 - modified by HC-2; 4 – modified by HC-3; 5 – modified by HC-4; 6 – modified by HC-5; 7 – modified by HC-6; 8 – modified by HC-7.

During addition of HC-3 and HC-4 such effect is not observed, probably these additives have inhibiting effect on thermal and thermooxidative destruction, as these additives do not have developed chain of conjugation. Taking into account the fact that polymer materials must work in narrow temperature ranges for a long time, we have studied also kinetics of thermal destruction at isothermal heating.

Data of kinetics of mass loss at samples warming up in the air at 200°C are given in Figure 3.18, from which it is seen that introduction of hexaazocyclanes into PETP decreases mass loss of polymer. Since for the period of 5 hours unmodified PETP – fibre loses about 5%, while PETP – fibres, modified by hexaazocyclanes, lose weight much less under the same conditions. For example, the fibre, modified by the dye HC-2, loses 0,8% of the weight. As at differential – thermal analysis HC-3 and HC-4 additives cause the increase of the rate of polymer mass loss.

From Figure 3.19 it is seen that introduction of hexaazocyclanes increases thermal stability of PETP in the air, moreover it depends on the chemical structure of additive and, specifically, on the degree of conjugation in molecule. As it is seen mass loss of unmodified sample increases almost linearly with the increase of temperature up to 390°C, while inhibiting effect in modified samples is observed up to 400-450°C, HC-3 and HC-4 being an exception. At much higher temperatures decay depth sharply increases.

Analysis of DTA, DTG, TG curves of modified and unmodified PETP – fibres has shown, that though all additives – modifiers being used belong to one class of compounds, they have different effect on PETP – fibre. Additives HC-2 and HC-1, being isomers of phenyldiamine (HC-1 – para-product; HC-2 – meta-product), have different effect on PETP – fibre. Additive HC-2 has greater plasticizing effect, then HC-1 (Table 29), though they have small difference in structure.

Introduction of HC-1 into polymer increases temperature of polymer decay at isothermal heating both in the presence of oxygen and without it. Additive HC-2 increases temperature of PETP decay still more (almost by 30°C higher, then HC-1). Hence, HC-2 protect PETP – fibre from thermal and thermooxidative destruction more effectively.

Results of introduction of HC-3 and HC-4 are depart from general tendency of hexaazocyclanes effect on PETP – fibre.

Introduction of HC-1, HC-2, HC-5, HC-6, HC-7 additives into PETP causes the increase of polymer resistance to thermal and thermooxidative destruction. Addition of HC-3 and HC-4 into PETP – fibre causes decrease of polymer resistance to these types of destruction.

Thus, all used additives may be devided into two groups: 1) increasing polymer resistance to thermal and thermooxidative destruction; 2) decreasing polymer thermal stability. Moreover, effect of additive on thermo- and thermooxidative stability will depend on the length of conjugation chain in modifier's molecule.

Irradiation of PETP – fibres was carried out in the air for evaluation of dyes effect on light stability of these fibres modified by hexaazocyclanes additives. Relative conservation of initial specific viscosity was taken as criterion of ultra-violet irradiation effect.

Table 30. Change of specific viscosity of PETP solutions after ultra-violet irradiation for 24 hours

Additive and its concentration in PETP, in mass %	Specific viscosity		
	Before irradiation	After irradiation	Conservation, %
Without additive	0,279	0,199	64,3
HC-1/1,0	0,335	0,241	71,8
HC-2/1,0	0,319	0,266	83,4
HC-3/1,0	0,312	0,194	61,5
HC-4/1,0	0,327	0,209	63,9
HC-5/1,0	0,325	0,236	72,6
HC-6/1,0	0,305	0,252	82,6
HC-7/1,0	0,346	0,256	74,0

Results of evaluation (Table 30) show, that little decrease of viscosity characteristics takes place at irradiation in the air, but besides specific viscosity of fibres is more conserved. HC-3 and HC-4 are the exception.

Photodestruction of PETP was studied under the action of ultra-violet light with different wave length 254 and 313nm. Intensity of incident light was 28 J/m^{2}c. Samples were prepared from initial and modified by HC-1 and HC-2 (1% of mass) PETP – fibres.

Under service conditions polymer materials are often subjected to simultaneous action of irradiation and other external factors in stressed state. Study of these complicated cases of destruction, when mechanically loaded polymer material is additionally subjected to the action for such external factor as ultra-violet irradiation, is considered to be important – scientific and practical – problem in physics and chemistry of polymers [304].

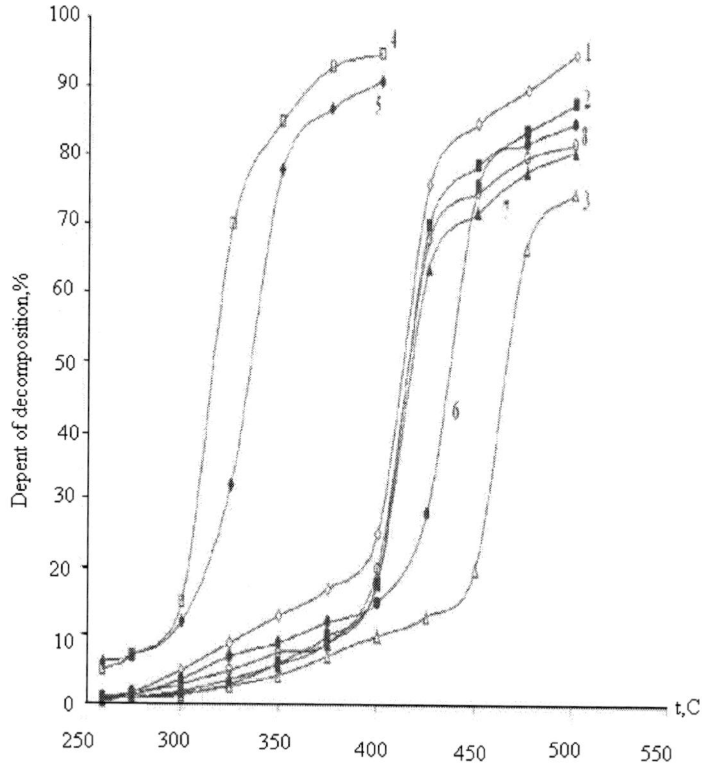

Figure 3.19. Kinetic curves of PETP – fibres decay in the air: 1 – unmodified; 2 – modified by HC-1; 3 – modified by HC-2; 4 – modified by HC-3; 5 – modified by HC-4; 6 – modified by HC-5; 7 – modified by HC-6; 8 – modified by HC-7.

In the work [305] it is shown that the load limits possible set of means of polymer stabilization and a number of light stabilizers, used for stabilization of non-loaded polymer samples, appeared to be low-effective for light stabilization of stressed samples. In connection with this light stabilizing effect of hexaazocyclanes has been studied not only under conditions of photoageing, but also of photomechanical destruction of PETP.

Results of the effect of preliminary ultra-violet irradiation by the light with different wave length on strength characteristics of initial and modified samples are given in Figure 3.20. From the figure it is seen that breaking strength of initial and ,modified samples without irradiation is different, however, comparison of results is not clear enough. For more evidence dependence of P on irradiation time for initial and stabilized samples, given in Figure 3.21, has been constructed according to these data, using formula:

$$P = (\delta_0 - \delta) / \delta_0 \bullet 100\%$$

where δ_0 and δ – breaking strength of initial and irradiated samples, P – degree of radiation damage. From the Figure 3.21 it is seen that HC-2 additive possesses light stabilizing effect at irradiation by ultra-violet light with wave length 254 nm and 313 nm.

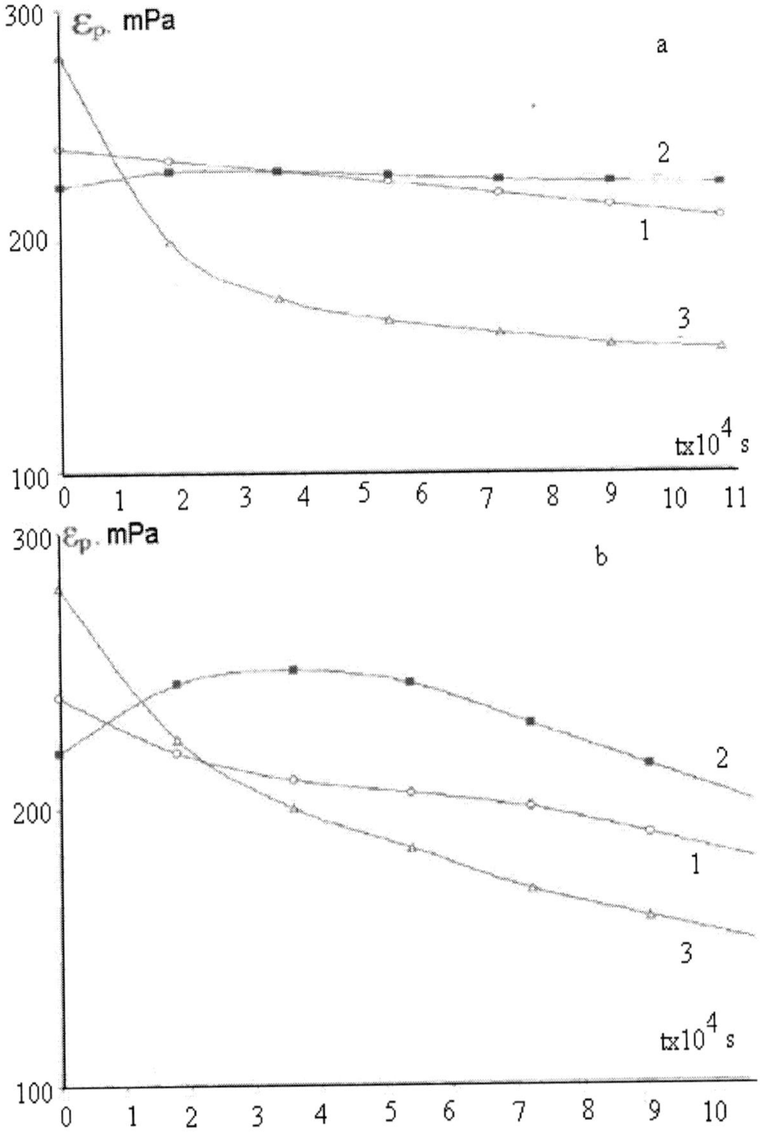

Figure 3.20. Dependence of breaking strength of initial and stabilized PETP – fibres on the time of radiation exposure: a) – λ=254 nm, b) – λ=313 nm; 1 – initial PETP; 2 – PETP + 1% HC-2; 3 – PETP + 1% HC-1.

Samples, containing HC-1 additives, break at a greater rate in comparison with initial samples, that shows sensibilizing action of HC-1 additive for irradiation by the light with wave length 254 nm and 313 nm. For example, after being irradiated by ultra-violet light with wave length 254 nm for 30 hours the degree of radiation damage in initial samples was only 12% while in those containing additive HC-1 it was 46%.

Thus, investigation of initial and stabilized PETP samples shows that additive HC-2 effectively stabilizes PETP at ultra-violet irradiation by the light with wave length 254 nm and 313 nm, and additive HC-1 acts as sensibilizer, speeding the process of PETP photodestruction at these wave lengths.

Effect of stabilization by hexaazocyclanes proved to be efficient at simultaneous action of both mechanical load and ultra-violet light. Dependence of durability of initial and dyed PETP samples on the value of load (Figure 3.22) at photomechanical destruction shows that durability of stabilized sample is 1,3 times higher than in initial one.

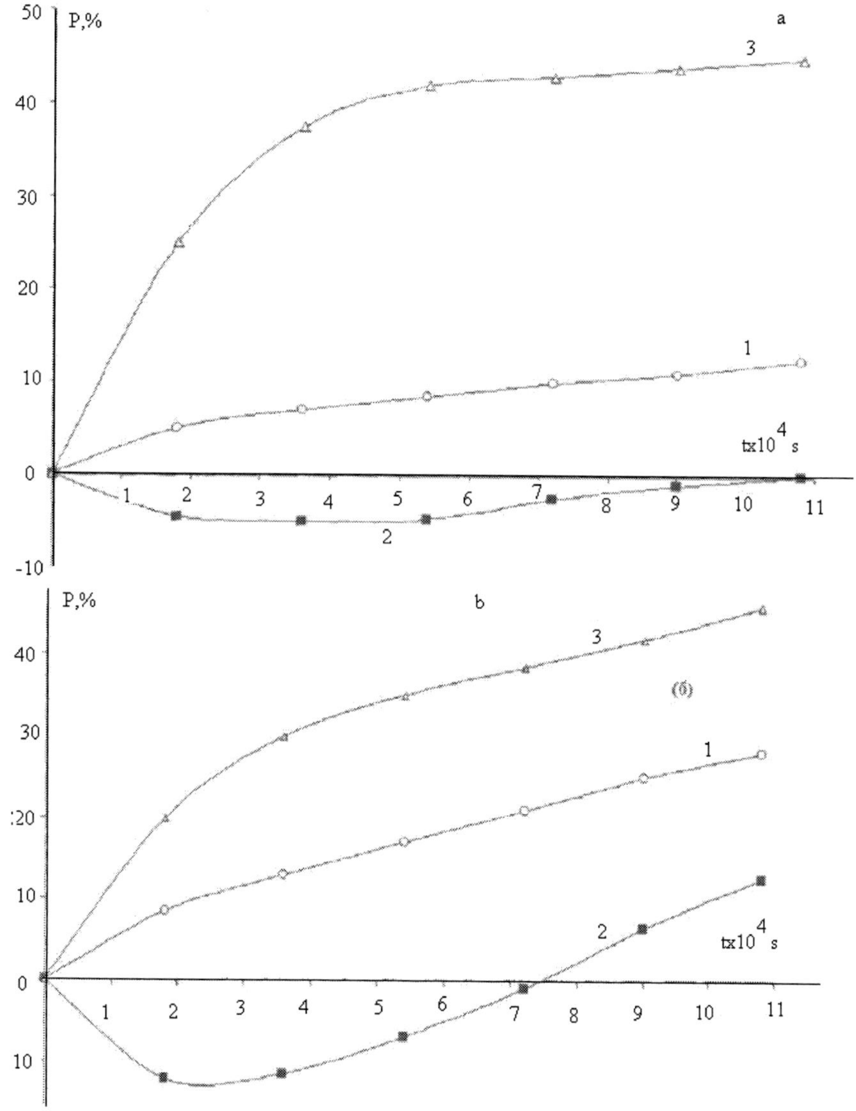

Figure 3.21. Dependence of breaking strength of initial and stabilized PETP – fibres on the irradiation time: a) – λ=254 nm, b) – λ=313 nm, 1 – initial PETP; 2 – PETP + 1% HC-2; 3 – PETP + 1% HC-1.

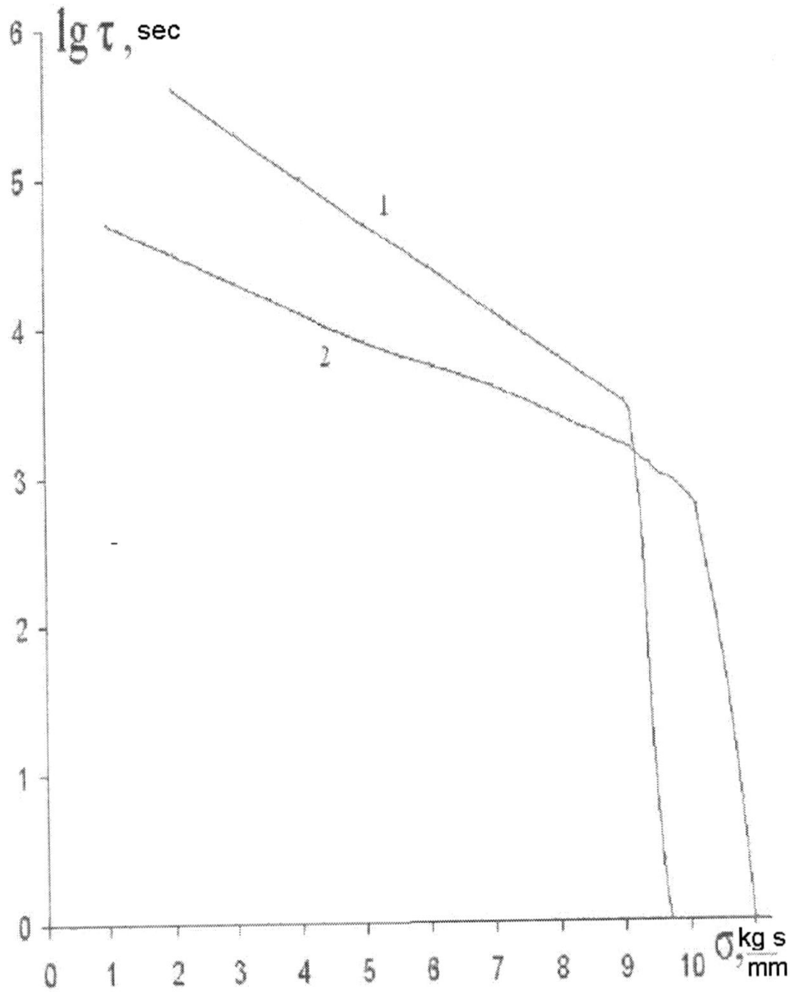

Figure 3.22. Dependence of durability of initial (1) and dyed (2) samples on the value of load.

Experimental data show that efficiency of light stabilization at photoageing depends on the degree of conjugation of hexaazocyclane being introduced into PETP.

On the basis of experimental data one may suppose that effect of light stabilization is probably connected with the rising of electronic excitation transfer from polymer to the additive. Besides, it may be expected that the more coplanar is the molecule, the more fluctuating degrees of release it has and the more is the observed effect of excitation suppression [306].

Important property of hexaazocylances is their ability to dyeing PETP into yellow, red, brown colours as in this case there is no need in additional introduction of the dye.

One of the features of consumer properties of chemical fibres and goods on their basis is light stability of polymer itself and its colour and also colour resistance to the action of wet treatment.

As a result of obtained PETP – fibres testing it has been found that colour resistance to light is 4-5 marks and to wet treatment it is 5/5/5, where the first index is colour change, the second – dye transfer into undyed PETP sample, the third – dye transfer into the matrix – viscose fibre.

Since operation of polymer goods is carried out as a rule, under conditions of variable humidity of environment; it was necessary to evaluate sorption properties of obtained PETP – fibres. In our case to describe sorption processes the following equation was chosen [307, 308]:

$$\alpha = \alpha_0 \exp [(-RT \ln \ln/E)' - \alpha (T\text{-}T)] \quad (5)$$

where α – value of equilibrium sorption, α_0 – maximum sorption from vapour phase, R – universal gas constant, T – temperature K, h – relative pressure of sorbate vapours, E – characteristic energy of absorption, T_0 – temperature K, at which constants of the equation were defined.

Choice of this equation is caused by the fact, that it satisfactorily described sorption isotherms not only of water, but of organic substances with root-mean-square error $\delta<13\%$, and it was successfully used for calculating sorption isotherm in systems PETP – methanol, PETP – phenol, PETP – water.

As sorption characteristics of irradiated and unirradiated dyed PETP – fibres, calculated from isotherms of water sorption (Table 31), show, sorption of water by dyed samples is higher than in initial ones, that may be connected with plasticizing action of hexaazocyclanes additives and also with the effect of additives on the character of mobility in PETP chains [309, 310].

Maximum sorption of irradiated fibres is considerably lower which is probably caused either by polymer construction seal in the process of irradiation, similar to the seal at thermal treatment, or by structurization taking place at severe irradiation and being similar to structurization in other polymers, namely, polystyrene [311].

All studied light stabilizers (Figure 3.1.) have >C=N- group. Photochemical reactions of >C=N- groups are similar to photochemical reactions of carbonyl groups [46]. That is why primary photochemical reaction of stabilizer should be breakage of H atom from substrate by excited state of >C=N- group.

Rate of deactivation of excite4d state for all HC – stabilizers being studied is much greater than the rate of intramolecular proton transfer.

It has been noted lately [14] that within the limits of one class of stabilizers change of their structure, introduction of substituents weakly change their reaction ability in solid polymer. Obtained by us results do not agree with this supposition, as HC-1 and HC-2 have different effect on thermal and thermooxidative stability of PETP – fibres.

Table 31. Value of sorption for irradiated and unirradiated PETP samples

Dye	Unirradiated		Irradiated	
	α_0, %	E, kJ/mole	α_0, %	E, kJ/mole
Without dye	1,41	1,60	-	-
HC-1	3,74	2,60	0,97	2,38
HC-2	2,70	1,53	1,54	3,53

It is known that more effective stabilizers are substances, possessing the largest system of conjugation. The whole complex of chemical and physical properties of polyconjugated systems is caused by special electronic structure of polyconjugated systems and, first of all, by delocalization of π – electrons along the units of conjugation chain. Necessary condition providing the greatest degree of delocalization of π – electrons, and hence, efficiency of conjugation, is coplanarity of fragments, composing conjugation system. That is more effective conjugated systems have joints whose space construction of a molecule approaches the plane.

While comparing structural formulae of modifiers and DTA, DTG, TG curves dyed by these additives of PETP – fibres it has been seen that effect of the dye on PETP – fibre is in direct dependence on structure and space configuration of dye molecules (hexaazocyclanes). Disturbance of coplanarity of heteroatom ρ – electrons, taking part in conjugation with π – electrons of the rest of the system, loads to the change in absorption spectra. At turning around simple bond in conjugated system by the angle less or equal to 45°C there occurs decrease of intensity of maximum absorption band without considerable shift λ_{max}. This effect is observed in absorption spectra of HC-1 and HC-2 (Figure 2). In absorption spectrum of HC-1 decrease of intensity of absorption band is observed comparing with absorption spectrum of HC-2. This has given the possibility to make a conclusion that a molecule of HC-2 has more plane structure than HC-1.

Computing of space configuration of hexaazocyclanes molecules (Figure 3.23 – 3.29) shows that the structure of HC-2 molecule is more planar (Figure 3.24) than of HC-1 (Figure 3.23). So, HC-2 possesses greater system of conjugation than HC-1 and, therefore, has more stabilizing effect. A great difference, being observed between values of A for HC-1 and HC-2, despite small difference in their chemical structures, is described by the fact that HC-2 structure is more planar than HC-1.

Plane structure of HC-2 must facilitate molecules aggregation of light stabilizer, that leads to very effective suppression of excited states. Computing of space configuration of hexaazocyclanes molecules also explains difference of properties of PETP – fibres and DAC films dyed by HC-3 and HC-4.

As it follows from abovementioned data, decrease of photosensiblizing HC action should improve its protective effect. That is why the problem of the effect of HC – light stabilizers structure on their photosensibilizing activity is very important.

Though molecule of HC-3 additive has the form close to the plane (Figure 3.25), its form is not a completed cycle (Figure 3.1) and this facilitates decrease of conjugation system. Molecule of HC-4 additive is not at all in two perpendicular planes (Figure 3.26), that is it is not conjugated. That is why, stabilizing action of these compounds is not observed.

Space configurations of molecules of the remaining hexaazocyclanes have the form close to plane (Figure 3.27 – 3.29).

From all hexaazocyclanes, being used, HC-2 has more plane molecule and, hence, greater degree of conjugation. Proceeding from this, HC-2 additive must possess the best stabilizing effect that is proved by complex thermogravimetric analysis and also by investigations of light destruction of DAC films modified by HC.

Figure 3.23. Structural formula of HC-1.

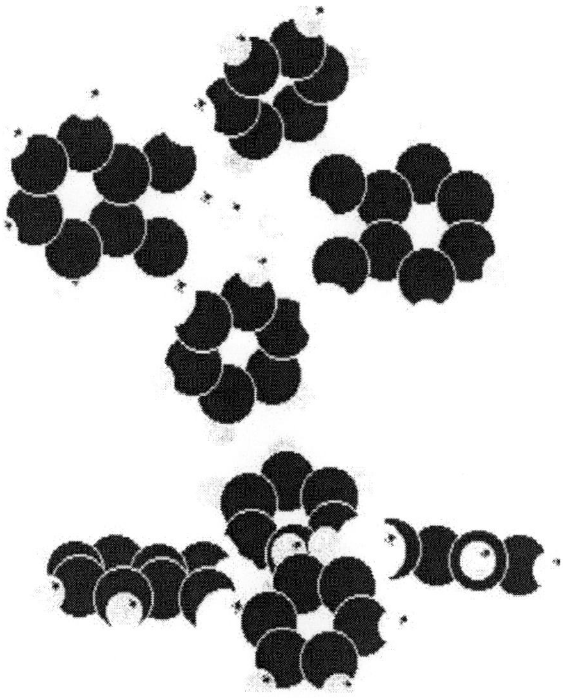

Figure 3.24. Structural formula of HC-2.

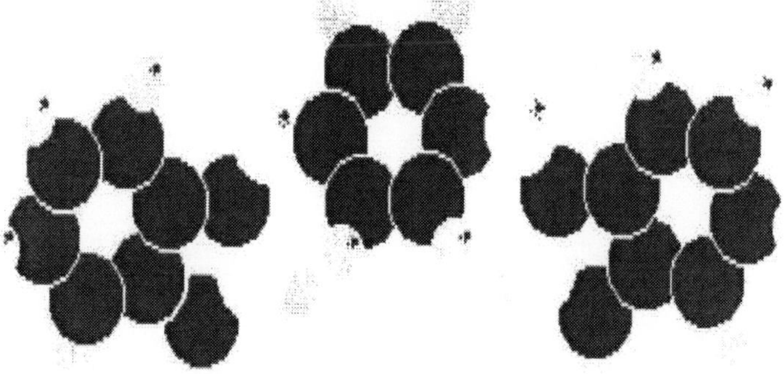

Figure 3.25. (Continued on next page).

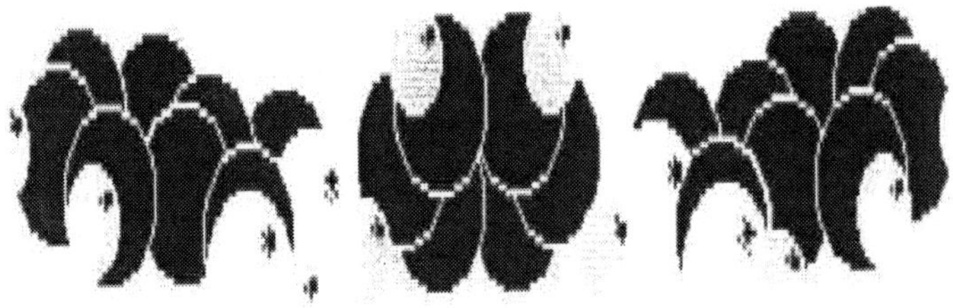

Figure 3.25. Structural formula of HC-3.

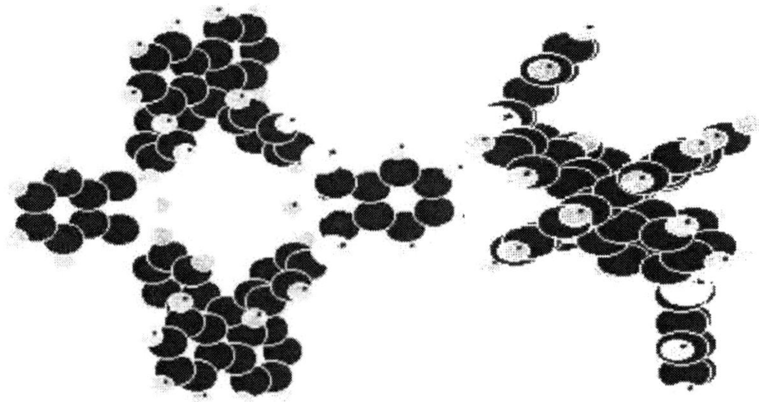

Figure 3.26. Structural formula of HC-4.

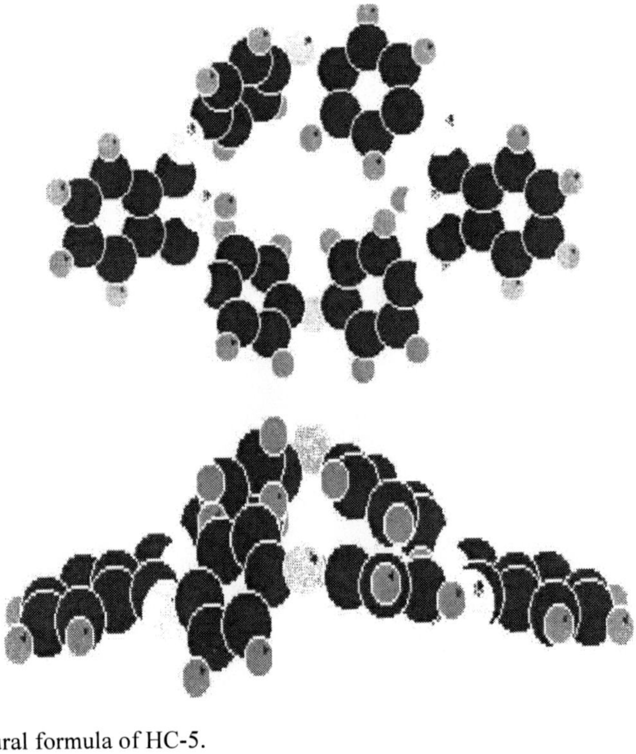

Figure 3.27. Structural formula of HC-5.

Figure 3.28. Structural formula of HC-6.

Figure 3.29. Structural formula of HC-7.

Catalysis of the reaction of kinetic chains transfer and freak is important property of solid-phase reactions and is of great importance in the processes of ageing and stabilization of polymers. Lowmolecular products of oxidation and destruction, dissolved in solid-phase polymer may react with macromolecules. Lowmolecular radicals being formed easily diffuse and, reacting in the other place, transfer free valency along the polymer volume, transmitting kinetic chains according to diffusive mechanism. Since diffusion of macroradicals takes place much easier in amorphous and intercrystalline regions of polymer (lowmolecular products of destruction are mainly localized there), then chain transfer by macroradicals defines, to a considerable extent, that fact that processes of oxidation and destruction of oxidation and destruction are localized mainly in these regions [312]. Hence, reduction of the number of amorphous sectors will prevent diffusion of macroradicals in polymer volume and kinetic chains will break. That is why destruction of polymer will be moderated. In microphotographs, presented in Chapter 3, it is seen that introduction of hexaazocyclanes decreases the number of amorphous regions (especially HC-2), that is why modification of PETP – fibres by hexaazocyclanes much cause decrease PETP resistance to light action.

Obtained results give the grounds to think that search of stabilizers for PETP should be carried out among macrocyclic aromatic compounds having structure approximating to a plane. Investigation of frequency dependences of dielectric characteristics has shown (Figure 3.30) that maximum of dielectric losses of initial PETP is in the region 1kHz, and in modified samples it slightly shifts into the region of higher frequencies and decreases by absolute value.

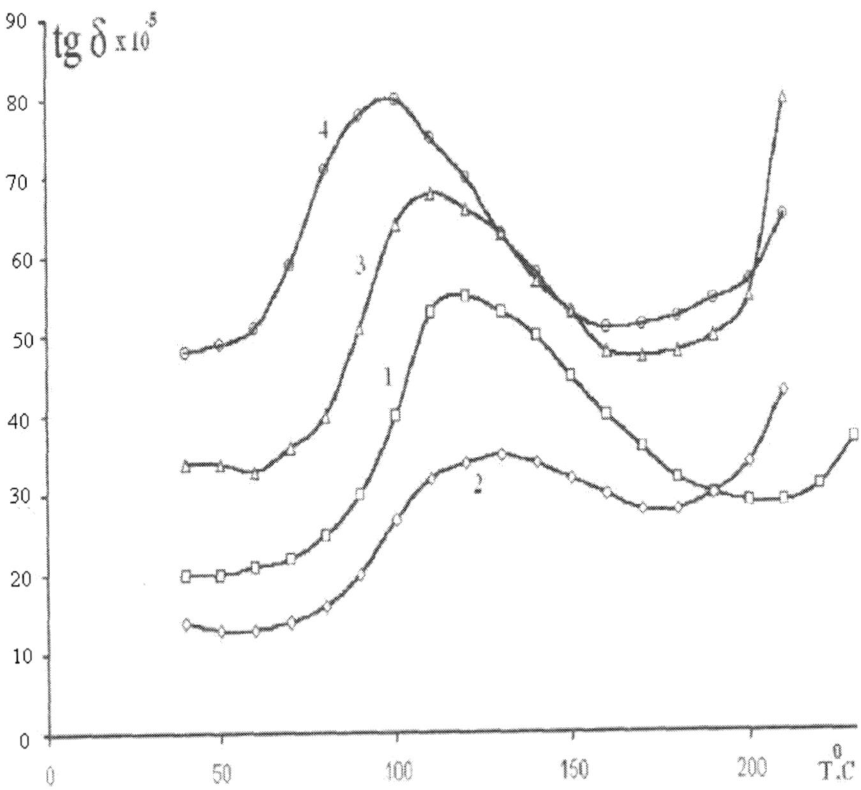

Figure 3.30. Dependence of tangent of dielectric losses angle on temperature of initial and modified PETP – samples: 1 – initial PETP; 2 – PETP + 0,1% of HC-2; 3 – PETP + 0,5% of HC-2; 4 – PETP + 3,0% of HC-2.

For initial PETP maximum of tangent of dielectric losses (tg δ), corresponding to dipole segmental mobility, is observed at 115°C. Position of this maximum on temperature axis during introduction of 0,1% of hexaazocyclane, does not change, and by absolute value it decreases by 1,5 times. Further growth of modifier concentration by 1-3% causes sharp decrease of tg δ value and maximum shifting accordingly by 10 °C and 25 °C into the region of lower temperatures that points to plasticizing action of modifier.

Further investigations have shown that, in spite of slight deterioration of dielectric properties and increase of electric conduction, polymer modification causes considerable improvement of mechanical properties ($\delta_{destr.}$ and $\rho_{destr.}$ increase almost by two times), moreover much higher values of strength and deformation are observed in samples containing 0,5% hexaazocyclane. Considerably growth of modified fibres strength is explained by plasticing effect of additive, since plasticization facilitates uniform distribution of stresses oriented strengthening of polymer.

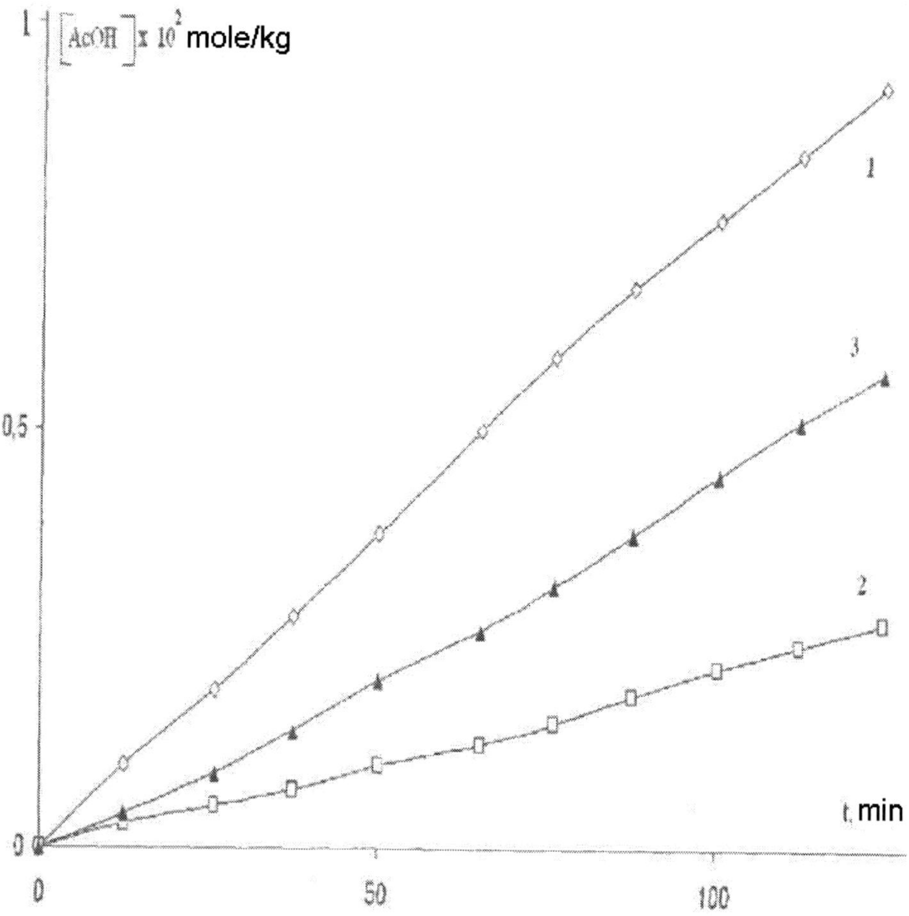

Figure 3. 31. Kinetic curves of acetic acid formation at irradiation of unstabilized CDA-films (1) and in the presence of GC – 1 (3) and GC – 2 (2) in vacuum at 298°K.

3.6. Light-Stabilization of Polyesters by Hexaazocyclanes

Energy, which molecule gets at photon absorption, may be consumed in photophysical and photochemical transformation.

Groups of atoms with double bonds absorb light with λ>250 in organic compounds, that do not contain other atoms except C, H, O, N. Energy levels, conforming these wave lengths, correspond to excitation energies of π-electrons with double bonds ($\pi \rightarrow \pi^*$ transition). Large wave lengths of incident light are sufficient for excitation of undivided electrons of heteroatoms, taking part in formation of double bonds with carbon atom ($n \rightarrow \pi^*$ transition).

First, valence bonds of macromolecule break under irradiation effect and radicals are being formed, which further undergo "dark" elementary reaction. For hemolytic break of the bond it is necessary for photon to possess definite minimum of energy, which depends on the wave length of photon.

In fact, energy, necessary for primary formation of radicals at polymer photolysis, is greater than the energy of bond dissociation, that is explained by the presence of cage effect in polymers [8].

Physical meaning of cage effect is in the fact, that a pair of free radicals, being formed at homolytical bond break, in condensed phase is surrounded by the "wall" of other molecules. Possibility, that two such radicals will recombinate before they will be able to disperse, is rather great. That is why additional energy is needed, and radicals, possessing this energy will be able to leave molecular cage. Besides particles, being in such cage, should be, in a definite way, oriented in respect to each other.

Restriction of molecular mobility is observed in solid matrices. Stabilization of not only free radicals, but instable molecular intermediate products may be the result of it.

Polyethyleneterephthalate (PETP) is resistant to ultra-violet irradiation, but at long sunlight action destruction becomes visible, especially during irradiation by the light with wave length from 300 to 330 nm [47]. Polyester does not destruct at photon energy lower than 3 e.v. (electron volt) [48].

Polyester fibre displays high stability at its protection from a part of spectrum, that causes its photochemical destruction.

In solid polymers, recombination of radicals is diffusive-controlling process (regardless of whether it is controlled by physical diffusion or chemical relay). High rate of diffusion leads to more uniform distribution of stabilizer in polymer mass and its fast migration to material surface, if necessary.

Probability of photooxidation is different in different points of polymer. First, particles, situated in the points, where quantum yield is higher undergo a reaction. That is why quantum yield gradually decreases during the reaction.

During photooxidation, limiting stage of the reaction $P^* + O_2 \rightarrow PO_2$ is microdiffusive stage, that is translational transport of dissolved oxygen before its meeting with macroradicals localized in lattice. Following kinetic stage – addition of oxygen to macroradical in the cage – is very fast.

During polymer oxidation, concentration of peroxide macroradicals and hydroperoxide groups should be the function of distance from polymer surface, even in that case when oxygen diffusion does not limit oxidation rate. For the same reason constant of the rate of radicals destruction should depend on the thickness of the sample. Constant of the rate should decrease to the value, characteristic for uncatalyzed recombinations, which is limited by the relay of valency migration, only in rather thin polymer films (-10^{-4} – 10^{-5}cm).

Thickness of absorbing layer should be as small as possible, as at $l \rightarrow 0$ the additive does not change quantity of light absorbed by polymer, that is effect of light diffusion processes through the polymer is eliminated. In optically thin layer $I_{abs.}$=2,3ecd.

In order to eliminate diffusion processes during investigation of light-stabilizing action of hexaazocyclanes, it is necessary to use films with small thickness as model object.

As it is rather difficult to obtain film from PETP, then films, obtained from cellulose diacetate (CDA), modified by hexaazocyclanes were used as samples.

Cellulose diacetate is related, as PETP, to polyesters, that is why effect of hexaazocyclanes on light-destruction of ester-groups may be investigated by using this material.

Photolysis of CDA runs with participation of two active centers: acetylalkyl radicals R and polyene radicals P', which differ hard in reaction ability. Acetic acid is formed as a result of CDA photolysis, that is why protective action of the additive may be judged by the quantity of extracted acid. Kinetic curves of acetic acid formation at irradiation of stabilized CDA films by the light with wave length of 254 nm. In all cases dependence of the amount of formed acid on the time of irradiation is a strait line from the tangent of slope angle of which the rate of acetic acid formation has been calculated. As it is shown in Figure 4.1, HC – 2 has strong protective effect on the rate of CDA - films photodestruction.

CONCLUSION

Thus, obtained results allow to consider that search for effective light-stabilizers should be carried out among macrocyclic aromatic compounds, possessing high degree of conjugation and having structure close to plane.

Sulfur- and nitrogen-containing hetrocyclic compounds I-IX, presented on pages, that quite well absorb photochemically active component of sunlight, owing to the system of conjugated bonds, have been studied as light-stabilizers. That is why they can "shield" hard ultra-violet irradiation and also may be "suppressors" of excited molecules of polymer, absorbed quantum of light. Presence of functional groups with labile hydrogen atom also assumes the possibility of these compounds action as inhibitors of oxidative destruction (antioxidants), and presence of chromophore groups shows the possibility of their use as polymer dyes.

Taking principles of "conjugation" as the basis, investigated compounds may be symbolically divided into four main groups (see appendex):

A – sulfur-containing heterocycles on the basis of benzo/v/thiophen (I-IX);
B – indan-carbazole-sulfoanilide (X-XXX);
C – Polyconjugated azomethine compounds (PAC) (XXXI-XXXVII);
D – naphthoilenebenzimidazoles and Hexaazocyclanes (XXXVIII – LVII).

APPENDIX

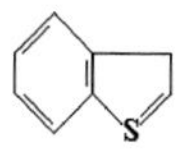

I

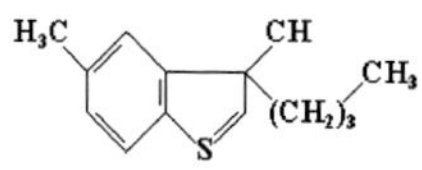

II

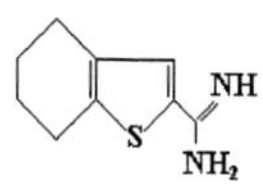

III

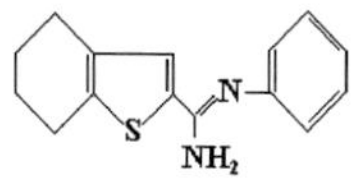

IV

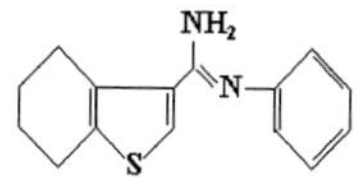

V

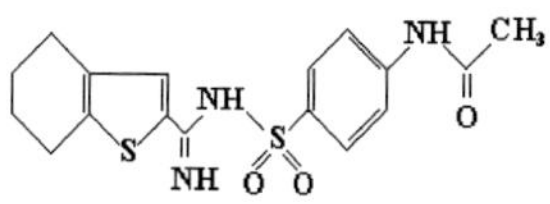

VI

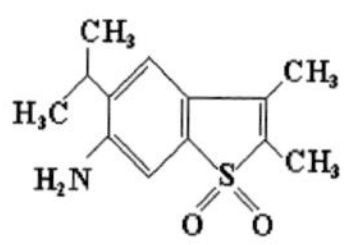

VII

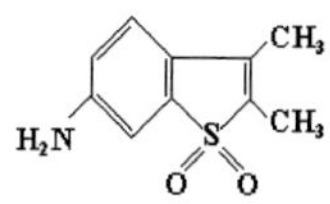

VIII

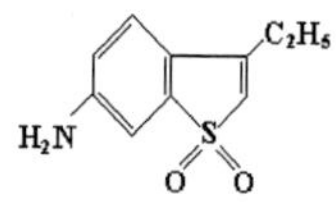

IX

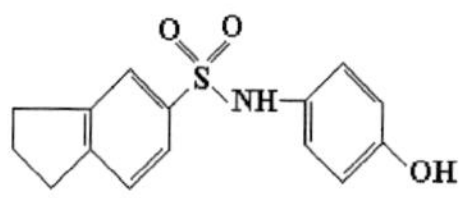

X

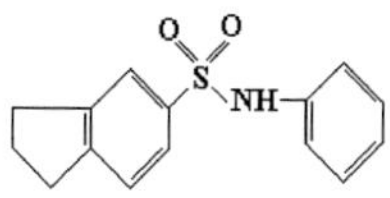

XI

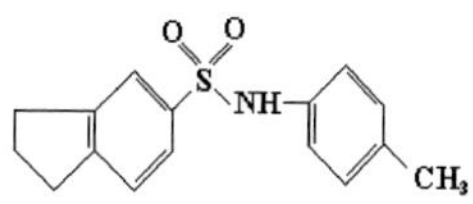

XII

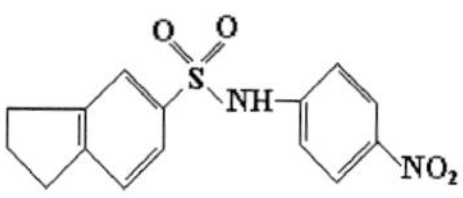

XIII

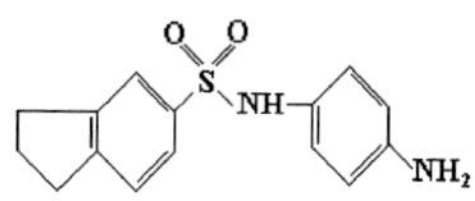

XIV

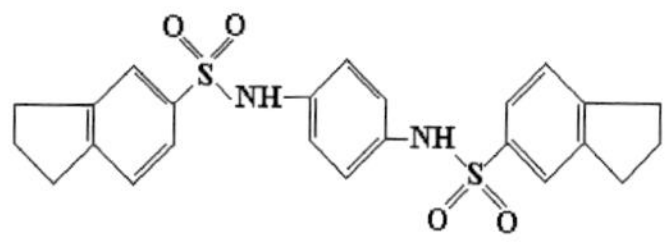

XV

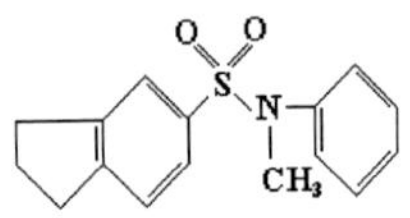

XVI

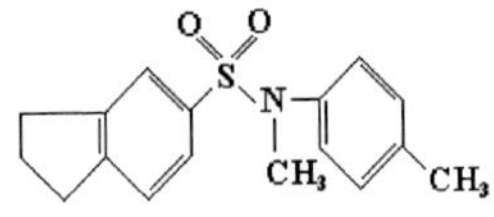

XVII

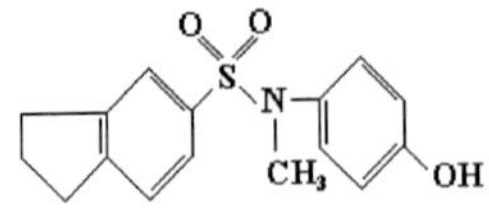

XVIII

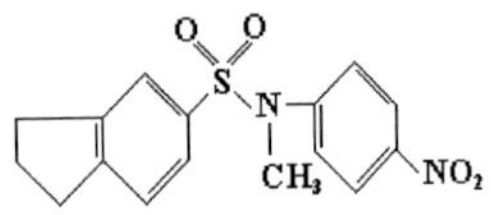

XIX

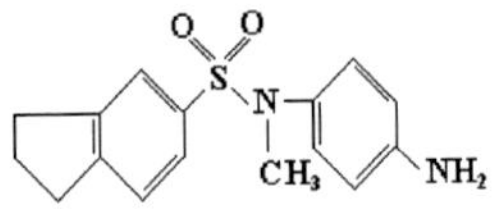

XX

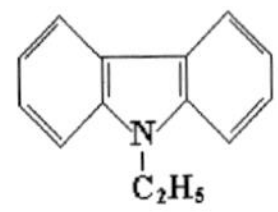

XXI

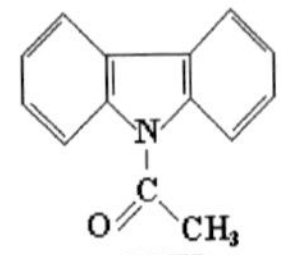

XXII

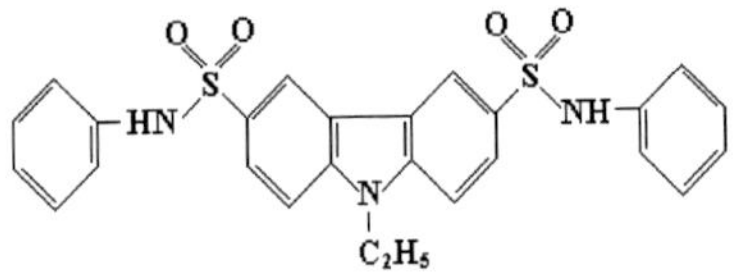

XXIII

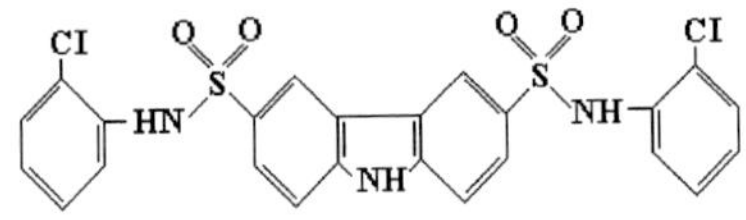

XXIV

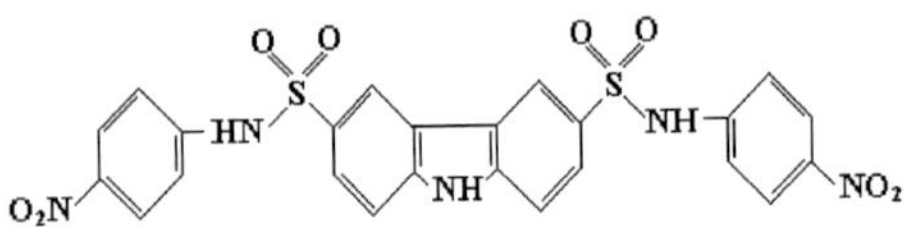

XXV

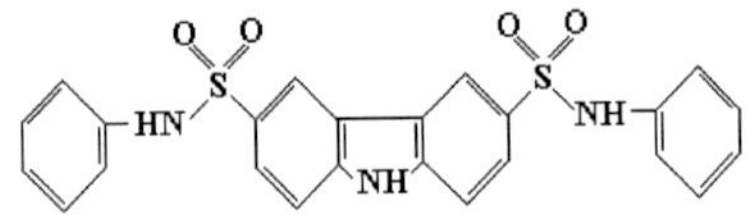

XXVI

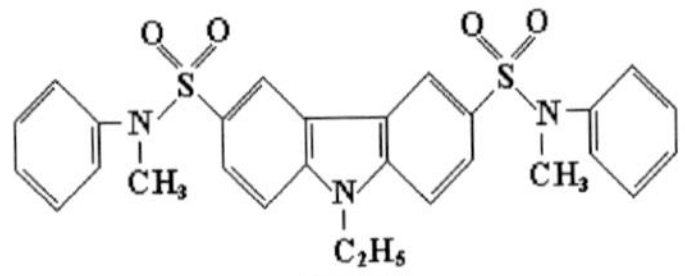

XXVII

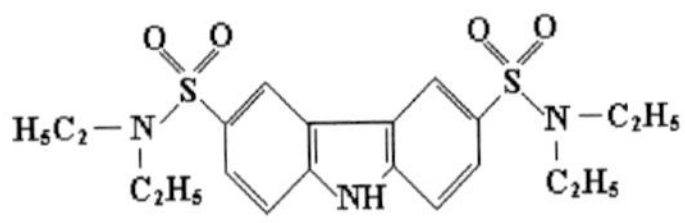

XXVIII

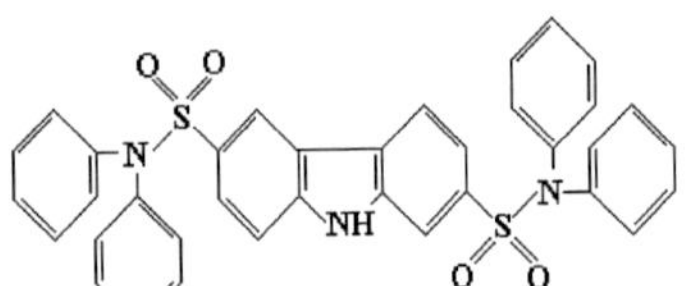

XXIX

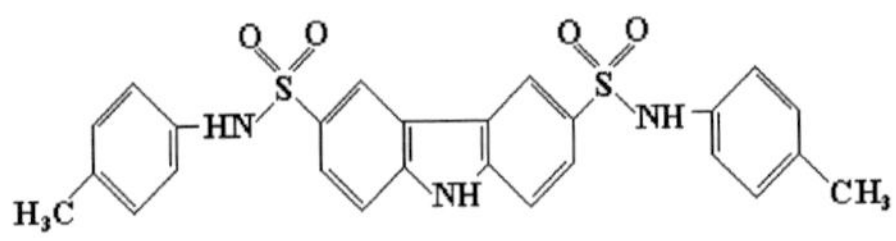

XXX

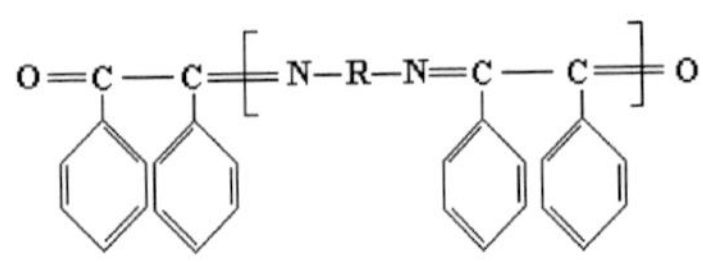

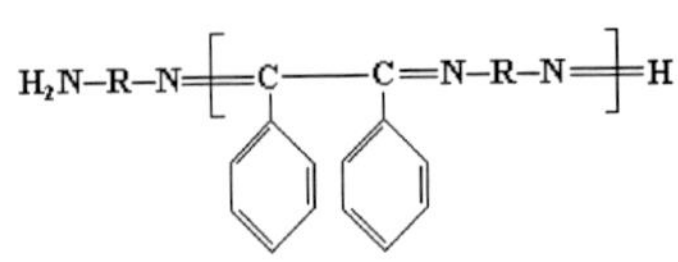

R = 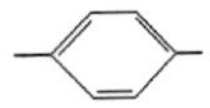 (M.M=400) - XXXI
(M.M=800) – XXXII

R = 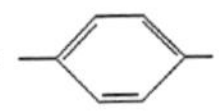 (M.M=600)-XXXIV
(M.M=1000)-XXXV

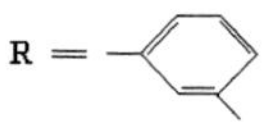

(M.M=1400) – XXXIII

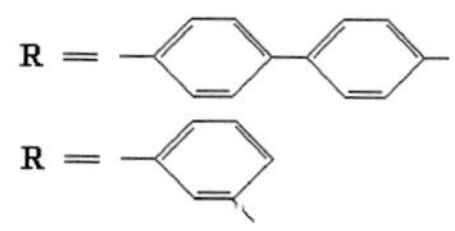

(M.M=1200)-XXXVI

(M.M=1680)-XXXVII

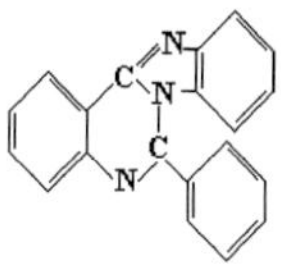

XXXVIII

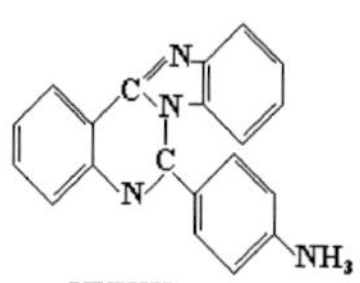

XXXIX

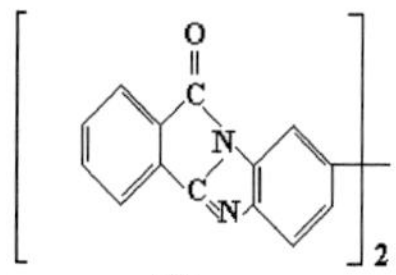

XL

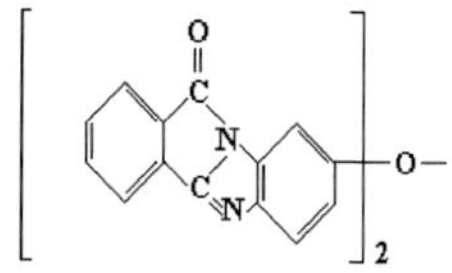

XLI

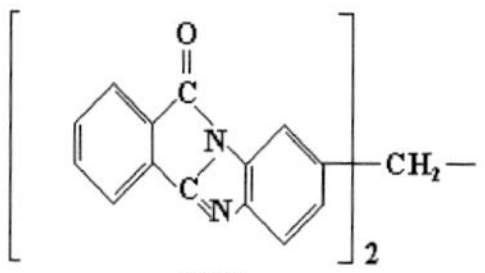

XLII

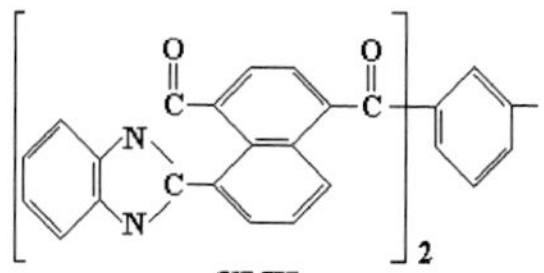

XLIII

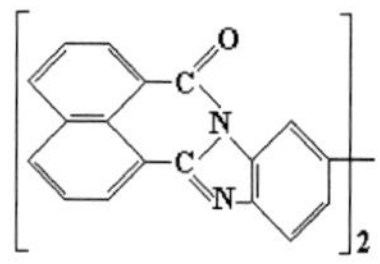

XLIV

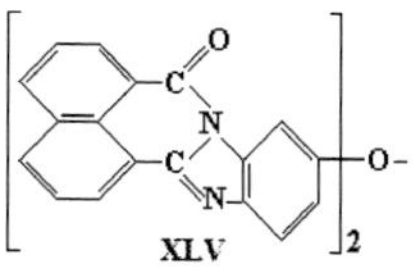

XLV

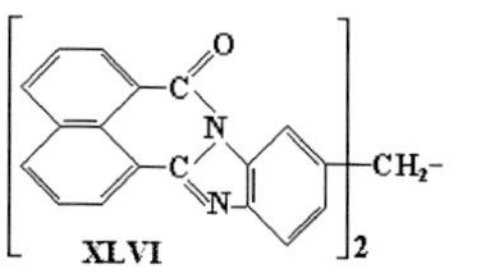

XLVI

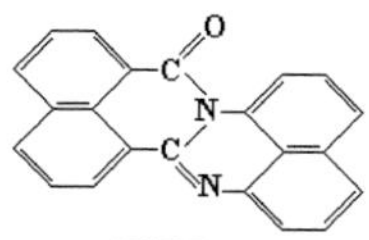

XLVII

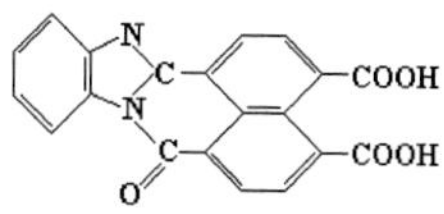

XLVIII

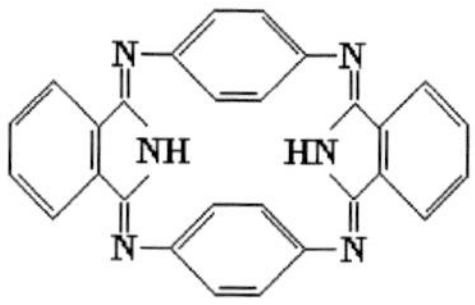

XLIX

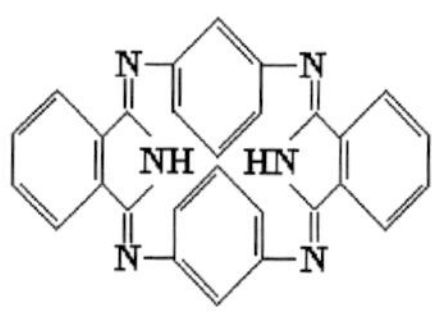

L

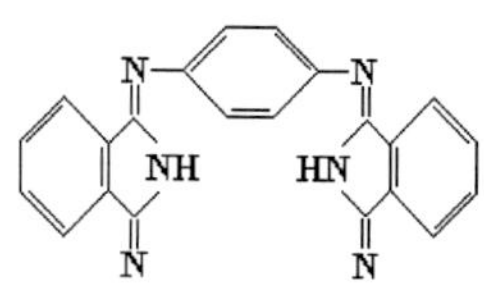

LI

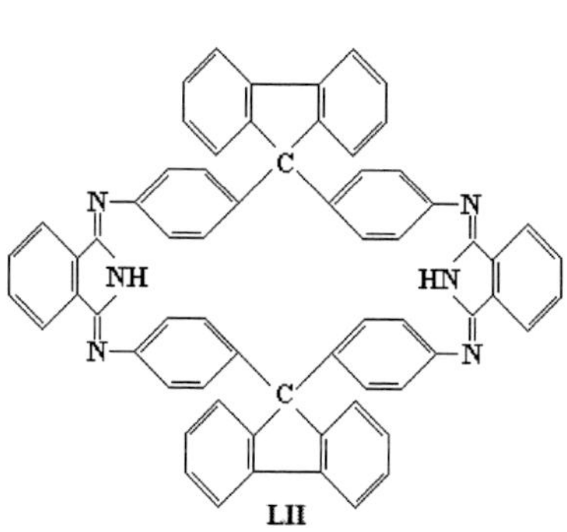

LII

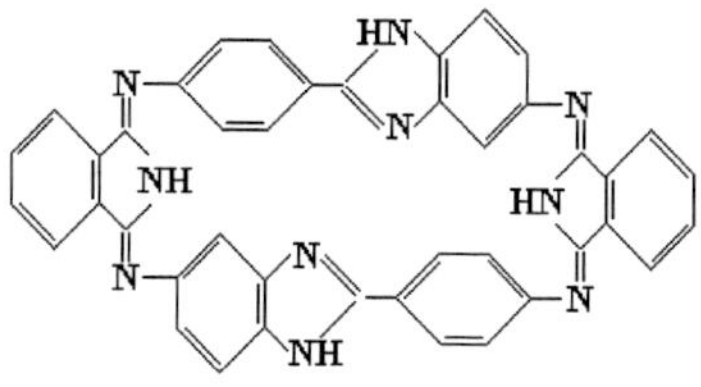

LIII

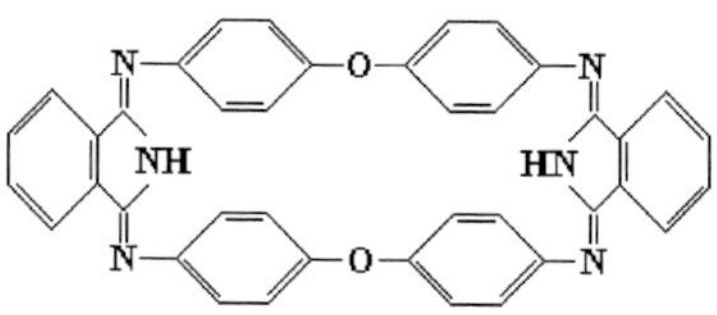

LIV

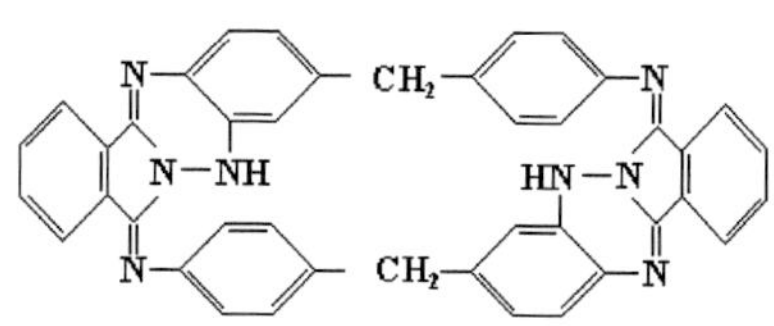

LV

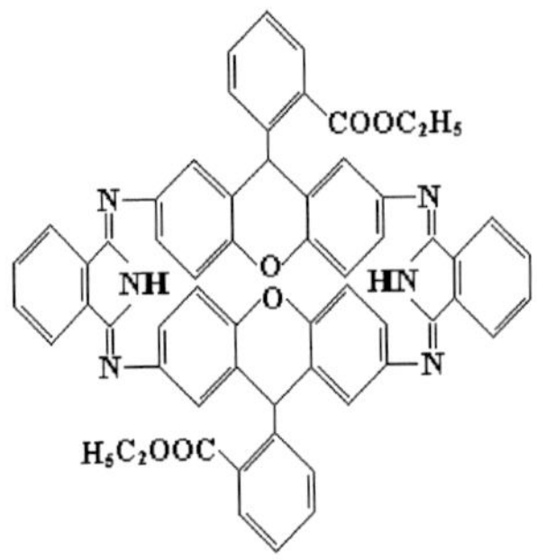

LVI

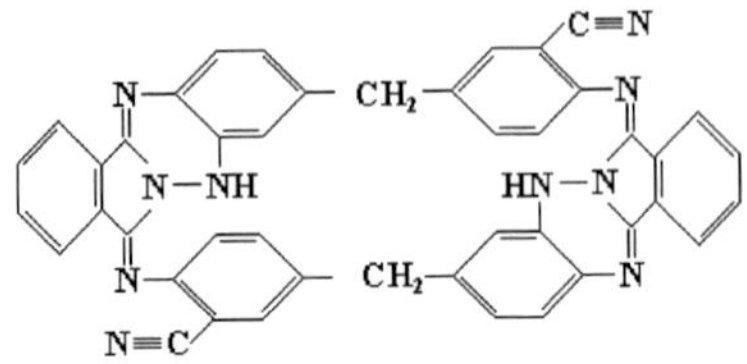

LVII

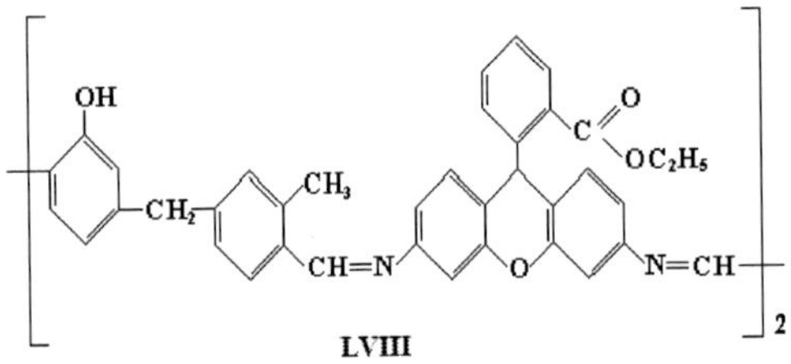

LVIII

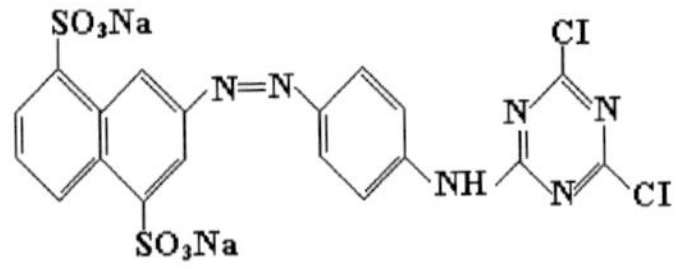

LIV

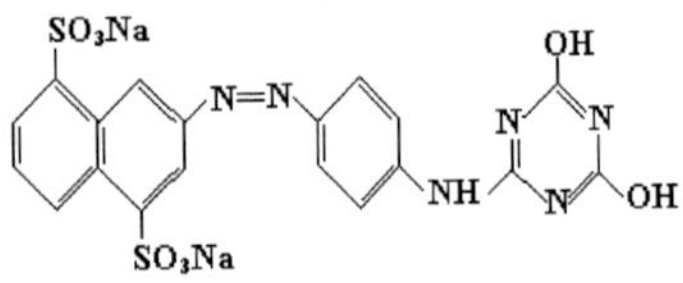

LX

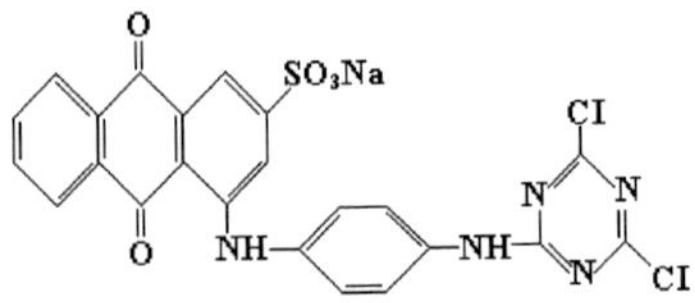

LXI

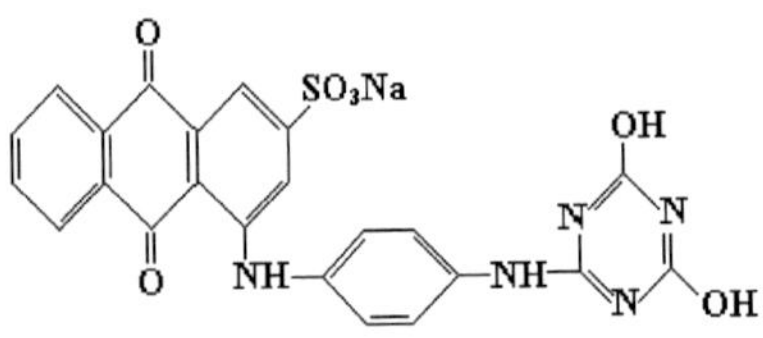

LXII

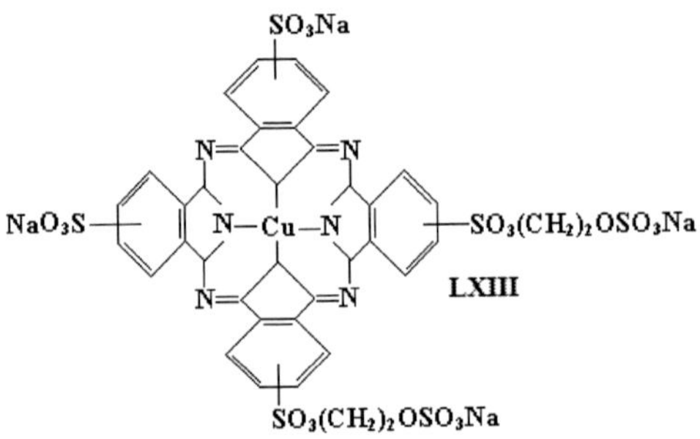

LXIII

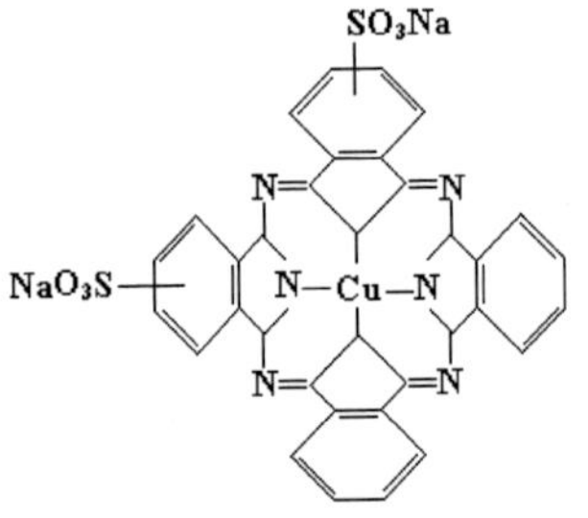

LXIV

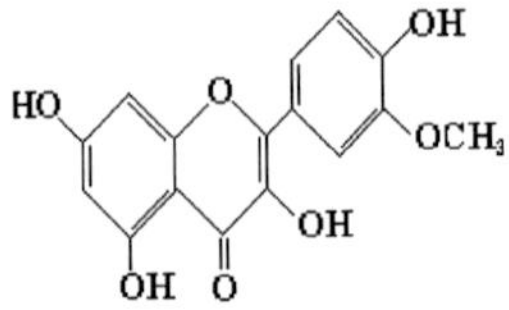

LXV

References

[1] J. Kalvert, J. Pitts, *Photochemistry*. Moscow, "Mir", 617 (*in Russian*).

[2] B. Renby, Ia. Rabek, *Photodestruction, photooxidation, Photostabilization of polymers*, Moscow, Publishing house "Mir" (1978), 676 (*in Russian*).

[3] A.L. Margolin, L.M. Postnikov, Photoageing of aliphatic polyamides//*Successes of chemistry* (1980), v. 49, No 6, 1106-1135 (*in Russian*).

[4] J.A. Dellingse, C.W. Roberts, Synthesis and luminescence properties of analogs of nylon-66//*J. Polym. Sci. Polym. Lett.* Ed. (1976), No 14, 147-148.

[5] N.S. Allen, Mc Reliar *Polymer//J.* (1975), v.7, 11-20.

[6] Ageing and stabilization of polymers//Editor Neiman M.E. „*Nauka*", Moscow (1969) (*in Russian*).

[7] H. Herlinger, B. Kuster, H. Essig, *Untersuchungen zum thermo-nud photooxidativen Abban von Polyamid- Textilien in Gegennart und Abwesenheit von Sticroxid*//Text. Prax int. (1989), 44, No 6, 655-556.

[8] S.R. Rafikov, *Ageing of polyamides*//Transactions of the Institute of chemical sciences of Academy of sciences of Kazakh SSR, Alma-Ata (1970), 3-39 (*in Russian*).

[9] D. Fromgeot, J. Zemaire, D. Sallet, Photooxydation de polyamides aliphatignes N-substitues par des gronpements alkyles//*Euz. Polymer. J.* (1990), 26, No 2, 1321-1328.

[10] V.Ia. Shliapintokh Photochemical transformations and light stabilization of polymers. Moscow: "*Chemistry*" (1979) (*in Russian*).

[11] L.M. Postnikov, *New mechanism of aliphatic polyamides photooxidation*//Report of Academy of sciences of the USSR (1985), v. 281, No 5, 1152-1156 (*in Russian*).

[12] A.L. Margolin, L.M. Postnikov, N.V. Semionova About some features of photooxidative destruction of polycaproamide//*Highmolecular compounds* (1974), A-16, No 5, 1037-1041 (*in Russian*).

[13] L.E. Nikulina, Effect of impurities of aniline, nitrobenzene and cyclohexanoloxime on caprolactam polymerization//*Chemical fibres* (1968), No 6, 15-17 (*in Russian*).

[14] L.E. Nikulina, Effect of impurities of caprolactam on photooxidation of kapron fibre//*Proceedings of higher schools. Chemistry technology* (1974), v.17, No 5, 778-780 (*in Russian*).

[15] A.L. Margolin, *On some features of photooxidative destruction of polycaproamide*//Dissertation of Candidate of chemical sciences, Moscow (1976) (*in Russian*).

[16] A.L. Margolin, Insensibilized photooxidative destruction of polycaproamide//A-17, *Highmolecular compounds* (1975), No 3, 617-620 (*in Russian*).

[17] Anton, Determination of Hydroperoxides in ultraviolet irradiated nylon 66//*J. Appl. Polym Sci.* (1965), v. 9, No 5, 1631-1639.
[18] Cicchetti, *Mechanism of Thermal Oxidation//Adv. Polym. Sci.* (1970), v. 7, 70-109.
[19] Ranby, J. Rabek, *J. Appl. Polym. Sci. Appl. Polym. Symp.* (1979), No 35, 243-263.
[20] E.V. Vichutinskaya, A.L. Margolin, L.M. Postnikov, V.Ia. Shliapintokh, Photooxidation of aliphatic polyamides under the action of long-wave ultra-violet light//*Highmolecular compounds* (1981), A-23, No 11, 2765-2771 (in Russian).
[21] E.V. Dovbiy, U.A. Ershov and others, Study of fluorence of polycaproamide//*Journal of applied spectroscopy* (1977), v. 25, No 5, 856-859 (in Russian).
[22] N.S. Allen, J.F. McKellar, C.B. Chapman, *J. Appl. Polym. Sci.* (1975), v. 20, 1717-1719.
[23] T. Karstens, V. Rossbach, Thermooxidative degradation of polyamide-6 and polyamide-6,6. Structure of UV/VIS-active chromophores//*Macromol. Chem.* (1990), v. 191, No 4, 757-771.
[24] H.D. Scharf, L-Ketoimidgruppierungen alscine der Ursachen fur die Lumineszene termooxidativ geschadigten Polycaprolactams//*Die Angewandte Macromolekulars Chimie* (1979), v.79, 193-206.
[25] L.M. Postnikov, I.N. Smolenskiy, E.V. Vichutinskaya, I.S. Lukomskaya, New mechanism of long-wave oxidation of polyamides//*Heads of the report. VI-th All-Union conference on ageing and stabilization of polymers*. Ufa (1983), 52 (in Russian).
[26] H.A. Taylor, Fotodegradation of nylon 6,6//*J. Appl. Polym. Sci.* (1970), v. 14, No 1, 141-146.
[27] G.E. Zaikov, Ageing and stabilization of polymers//*Chemistry successes* (1991), 60, No 10, 2220-2249 (in Russian).
[28] G. Clare, E. Fritsche, F. Grebbe, Synthetic polyamide fibres. Moscow, "*Mir*" (1966) (in Russian).
[29] T.S. Allen, J.F. McKellar, D. Wilson, Luminescence and degradation of nylon polymer. III-Eridance for triplet-singlet resonance energy transfer into dyes//*J. Photochem.* (1977), v.7, No 1, 405-409.
[30] G. Egerten, Fundamental principles of photochemistry of dyes//*J. Sci. Dyer. Coiur.* (1970), v. 86, No 2, 72-83.
[31] A.M. Terenin Photonics of dyes molecules and related organic compounds. Leningrad, "*Science*" (1967), 516 (in Russian).
[32] C.H. Bamford, Photosensibilization by Vat. Dyes//*Nature* (1949), v.163, No 1, 214.
[33] B.M. Povarskaya, A.B. Blumenfel'd, I.I. LEvantovskaya, Thermal stability of heterochain polymers. Moscow, *Chemistry* (1977) (in Russian).
[34] A.E. Neverov, G.A. Nikolaev, Study of kinetics of oxygen absorption in the process of polycaproamide thermooxidation and prediction of its storage life//*Highmolecular compounds* (1982), A-24, No 4, 272-275 (in Russian).
[35] P.M. Pakhomov, L.S. Gerasimova, S.A. Baranova, E.V. Shablygin, Change of the structure of polycaproamide mechanical properties at thermal ageing//*Highmolecular compounds* (1984), v. 26, No 2, 153-157 (in Russian).
[36] L.I. Shushkevitch, L.A. Krul', M.A. Matusevitch, Thermal analysis of the products of interaction of 1,4 benzochinone with polycaproamide//*Journal "Applied chemistry"* (1983), v. 19, 2104-2106 (in Russian).

[37] G.G. Makarov, L.M. Postnikov, G.B. Pariyskiy, U.A. Mikheev, D.D. Toptygin, Study of kinetics of thermal decay of benzoyl peroxide in polyamide and destruction of polymer macromolecules//*Highmolecular compounds* (1978), A-20, No 11, 2567-2572 (in Russian).

[38] U.M. Vasiliev, Study of products of polycaproamides thermal decay by the method of infrared spectroscopy//*Physics-chemistry of multiphasw systems*. Balashov (1980), 39-46 (in Russian).

[39] G.G. Makarov, Radical reactions of aliphatic polyamide destruction initiated by thermal and photodecay of peroxides//*Author's abstract of candidate dissertation*, Moscow (1983) (in Russian).

[40] G.G. Makarov, Yu.K. Mikheev, G.B. Pariyskiy, L.M. Postnikov, D.D. Toptygin, About mechanism of reactions initiated by thermal decay of benzoil peroxide in polyamide//*Highmolecular compounds* (1982), A-24, No 12, 2601-2508 (in Russian).

[41] N.S. Allen, Developments in Polymer Photochemistry//*Ed.Letter. Appl. Sci. Publ.* (1980), 223.

[42] Patent of England M987448, *Stabilized polymeric compositions. Claimed* – 12.01.1962. Published – 1966.

[43] R. Foigt, *Stabilization of synthetic polymers against light and heat action*, Leningrad (1972), publishing house "Chemistry" (in Russian).

[44] N.S. Allen, *Polym. Sol.. Polym. Chem*. Ed. (1974), v. 12, No 6, 1233-1241.

[45] Chemical additives for polymers. Moscow, *Chemistry* (1973), 11 (in Russian).

[46] L. Maskis, Additives for plastic mass. Moscow, *Chemistry* (1978) (in Russian).

[47] N.S. Allen, *J. Polym. Sci. Polym. Chem* Ed. (1975), v. 13, No 12, 2857-2858.

[48] P. Rochas, *Bull. Inst.Text*. France (1959), v. 83, 41-84.

[49] R.V. Subramanian, Photodegradation of Nylon 6//*Text. Res. J.* (1972), v.42, No 4, 207-214.

[50] IV International conference on achievements in the field of stabilization and control over the processes of polymer destruction//*Highmolecular compounds* (1933), A-35, No 2, 663-664 (in Russian).

[51] E.M. Emanuel' *Up-to-date state and perspectives of the development in the field of ageing and stabilization of polymers*. Vilnius (1980), 3 (in Russian).

[52] B. Ranby, I.O. Vog, U.I. Sweden Workshop on Photodegradation and Photostabilization of Polymers //*Polym. news* (1982),v. 8, No 5, 188-189.

[53] A.L. Margolin, G.S. Kabanova, L.M. Postnikov, V.Ja. Shliapintokh, Photooxidation of aliphatic polyamides//*Highmolecular compounds* (1976), A-18, No 5, 1094-1097 (in Russian).

[54] Patent of England No 652947 (R.ZhH, 1971, 22C413P) (in Russian).

[55] Claim for an invention, Japan, RZhH 4C215P (1972) (in Russian).

[56] R.P. Smirnov, V.M. Kharitonov, L.E. Smirnov *Transactions of chemical-technological institute*, Ivanovo (1972), issue 14, 111-113 (in Russian).

[57] V.M. Levin, V.K. Muraviev, L.E. Smirnov, *New stabilizing additives. III International symposium on chemical fibres*. Kalinin (1981), No 5, 288-197 (in Russian).

[58] Copyright certificate of the USSR No 322349. *Bulletin of inventions* (1971), No 36 (in Russian).

[59] Ja. Kalontarov, F.F. Niyazi, Yu.V. Chaiko, S.A. Siling, K.I. Ponomarenko Polyhexaazocyclenes as photo-thermo stabilizers of polymers. *Plastic masses* (1983), No 12, 49-50 (in Russian).
[60] J. Gallovic, V. Kvarda, Copyright certificate No 258747 of Chekhoslovakia (in Russian).
[61] E.J. Hagon, *Amer. Chem. Soc.* (1970), v. 91, No 13, 3893-3903.
[62] A.A. Berlin, Chemistry of Polyconjugated systems//Moscow, *Chemistry* (1972) (in Russian).
[63] De Cross, G.J. Tamblyn W. *Moel. Plast* 952, v.29, No 8, 127.
[64] Brawn Hs Ind. *Eng. Chem.* (1965), v. 57, 102-105.
[65] Light stabilizers of polymer materials. Collection "*Correlating surveys on separate productions of chemical industry*", Moscow (1971) (in Russian).
[66] A.L. Kazilinnas, I.E. Smolenskiy, A.P. Paulauskas, Light-stabilizing efficiency diaryloxamides derivatives at dyes fading and photodestruction of fibres. *Textile industry* (1982), No 9, 56-57 (in Russian).
[67] Chemical additives for polymers. Reference book under the editorship of Maslova I.I., Moscow, "*Chemistry*" (1981), 264 (in Russian).
[68] J. Zimmerman Spectra of irradiated polyamides//*J. Appl. Polym. Sci.* (1959), v. 2, No 5, 181-185.
[69] V.Ja. Shliapintokh General mechanisms of polymers light-stabilization. Heads of the report. V All-Union conference "*Ageing and light-stabilization of polymers*" Vilnius (1980), 9 (in Russian).
[70] C.F. Wells, Hydrogen Transfer to Quinones Trans.//*Faraday Soc.* (1961), No 57, 1705-1719.
[71] B.M. Kovarskaya, Bliumenfel'd Amine stabilizers of polyamides. *Plastics* (1965) No 8, 8 (in Russian).
[72] Copyright certificate of the USSR No 218415. *Bulletin of inventions* (1968) No 7 (in Russian).
[73] L.V. Kut'ina, *Synthesis and investigation of chemicals efficiency for polymer materials*. Tambov (1969), issue 3, 158-161 (in Russian).
[74] Copyright certificate of the USSR No 353943. *Bulletin of inventions* (1972), No 30 (in Russian).
[75] H.H. Dearman On the Photochemistry//*J.Chem. Phys.* (1966), v. 44, No 1, 416-417.
[76] E.G. Makhviladze, *Photofading and photodtabilization of dispersed dyes on polyamide fibres*. Author's abstract of candidate dissertation, Moscow (1981) (in Russian).
[77] N.V. Lysun, M.D. Kats, G.M. Krichevskiy, Directional synthesis of dyes with given properties regarding polyamide fibres. *Proceedings of higher schools "Chemistry and chemical technology"* (1984), 27, No 6, 700-703 (in Russian).
[78] G.E. Zaikov, Up-to-date state and perspectives of the development of investigations in the field of ageing and stabilization of polymers. *Plastic masses* (1991), No 5, 30-37 (in Russian).
[79] Ja. Kalontratov, A.E. Kharkharov, Effect of light weather on physico-mechanical properties of kapron fibre dyed by dichlorotriazine dyes. *Report of the Academy of sciences of Tadzhic SSR* (1967), 10 No 5, 50-52 (in Russian).

[80] E.V. Lysun, *Development of the method of polyamide textile materials stabilization by dispersed dyes*. Author's abstract of candidate dissertation, Moscow (1984) (in Russian).

[81] Margolin, A.V. Sorokina, I.M. Nosalcvitch, A.I. Pervykh, Effect of light-stabilizers on chain photooxidation of polyamide in the conditions of weak and strong light absorption. *Highmolecular compounds*. A (1983), v. 25, No 4, 771-775 (in Russian).

[82] E.V. Lysun, A.L. Margolin, G.M. Krichevskiy, Discovery of reasons of light-protective action of dyes regarding polyamide materials. *Proceedings of higher schools. Technology of textile industry* (1982), No 3, 62-66 (in Russian).

[83] V. Khabarov, Effect of dyes additives on radiation-chemical transformations of polyamide. *Chemistry of high energies* (1982), 16, No 1, 32-36 (in Russian).

[84] L.Yu. Kokinaite, A.L. Kaziliunas, Effect of modifier and azo dyes on kinetics of photolysis and photooxidative destruction of kapron fibres. Scientific transactions of higher schools of Lithuanian SSR. *Chemistry and chemical technology* (1980), No 8, 32-41 (in Russian).

[85] Ya. Kalontarov, The effect of Dyes on the degradation of fibreforming polymers//*Polymer Yearbook* (1991), v.7, 43-61.

[86] J. Senedirty, K. Dudrejmisra, Farbenie PAD-5 vlaken v hmote a farebna pruhovitost//*Polyamide hodvab*. L b prednas. Konf. Chemlon. n.p Humenne 5-6 sept 1984, Kosict, 56-99.

[87] L.P. Gritsenko, Yu.P. Tretiakov, Photochemical transformation of 1-acetoxy-3-methyl-anthrapyridones in polyamide substrate. *Ukrainian chemical journal* (1980), 46, No 7, 755-759 (in Russian).

[88] Claim for an invention. Japan No 56-27609. *Dyed polyamide fibre*. Published 25.06.1981 (in Russian).

[89] Patent of Czechoslovakia N0 121419. Published 16.03.1966 (in Russian).

[90] English patent No 964602. Published in 1966 (in Russian).

[91] English patent No 963783. Published in 1964 (in Russian).

[92] Patent of German Democratic Republic No 11939. Published in 1956 (in Russian).

[93] A.R. Shultz, *J. Appl. Polyin Sci.* (1966), 10, No 3, 353-359.

[94] English patent No 360628 (1961) (in Russian).

[95] K. Venkataraman, *Chemistry of synthetic dyes*. Moscow, Gostekhizdat (1956), v. 1, 11-14 (in Russian).

[96] Copyright certificate No 230235 of Czechoslovakia. *Method of polyamide fibres* dyeing (1987) (in Russian).

[97] Copyright certificate No 235160 of Czechoslovakia. *Dyeing of polyamide products in mass into yellow* (1987) (in Russian).

[98] Claim for an invention No 1903234. *FRG. Bulletin of inventions* (1975), No 5 (in Russian).

[99] Patent of the USA No 3878158. *Bulletin of inventions* (1975), No 12 (in Russian).

[100] English patent No 1004993. *Dyed polymer composition* (in Russian).

[101] Patent of the USA No 3388094 (1968) (in Russian).

[102] Swiss patent No 411333 (1968) (in Russian).

[103] M.V. Kazankov, V.E. Ufimtsev, Copyright certificate of the USSR No 197087. *Bulletin of inventions* (1968), No 33 (in Russian).

[104] Patent of FRG No1903244. *Method of synthetic linear polyamide dyeing in mass* (1978) (in Russian).
[105] Patent of FRG No 1303244. *Dyes for polyamide* (1978) (in Russian).
[106] Patent of FRG No 1215844 (1966) (in Russian).
[107] M.V. Kazankov, Dyes for synthetic fibres dyeing. *Journal of All-Union chemical society named after D.I. Mendeleev* (1974), v. 19, No 1, 64-71 (in Russian).
[108] Copyright certificate of Czechoslovakia No 163485 (1979) (in Russian).
[109] *Polyamide fibres resistant to photo-and thermodestruction*. Copyright certificate of Czechoslovakia No 193874 (1982) (in Russian).
[110] Copyright certificate of Czechoslovakia No 207878 (1984) (in Russian).
[111] *Thermo-and light-stabilizing polyamide fibre*. Copyright certificate of Czechoslovakia No 193882 (1982) (in Russian).
[112] P.Ja. Kostorina, Development of dyeing concentrates on polymer base for Kapron filaments dyeing in mass. *All-Union scientific-technical conference "Theory and practice of chemical fibres formation"*, Mytishchi (1933), 123-129 (in Russian).
[113] B.M. Krasovitskiy, K.F. Levtchenko, Organic luminophores. *Collection "Monocrystalls, scintillators and organic luminophores"*, Kharkov State University (1967), 90-94 (in Russian).
[114] N.F. Levchenko, Luminescent dyes for polymers. Collection "Monocrystalls, scintillators and organic luminophores. *Transactions of All-Union scientific-research institute of monocrystalls*, Kharkov (1968), 186-189 (in Russian).
[115] S.E. Kovaliev, Copyright certificate of the USSR No 327225 (1972). *Bulletin of inventions*, No 5 (in Russian).
[116] S.E. Kovaliev, Naphthoilenebenzimidazoles as luminophores for polymer materials. "*Chemistry of heterocyclic compounds*" (1974), No 4, 461-463 (in Russian).
[117] S.E. Kovaliev, Copyright certificate of the USSR No 327503 (1972). *Bulletin of inventions*, No 7 (in Russian).
[118] E.A. Shevtchenko, *Author's abstract of candidate dissertation*, Kharkov, All-Union scientific-research institute of monocrystalls (1970) (in Russian).
[119] L.L. Ostis Active luminescent dyes on the basis of 4,5-carboxy-1,8 naphthoilen-1^1,2^1-benzimidazoles. *Dissertation of the candidate of chemical sciences*, Kharkov (1979) (in Russian).
[120] A.M. Kuznetsov, Copyright certificate of the USSR No 309036, "Organic luminophores", Leningrad, "*Chemistry*" (1975), 210 (in Russian).
[121] B.M. Krasovitskiy, B.M. Bolotin Organic luminophores. Leningrad, "*Chemistry*" (1976), 210 (in Russian).
[122] V.L. Bell, Polyamidazopyrolons//*Encycl. Polym.Sci. and Technol.* (1969), v. 11, 240-246.
[123] V.V. Korshak, Copyright certificate of the USSR No 293016 (1971). *Bulletin of inventions,* No 5 (in Russian).
[124] Z.I. Kuznetsov, S.V. Savin, Obtaining of copolyaroilenebenzimidazoles of asymmetrical structure. *Journal of applied chemistry* (1959), v. 32, No 10, 23 (in Russian).
[125] V.V. Korshak, A.L. Rusanov, F.F. Niyazi, I. Batirov, Synthesis and investigation of some bis-(1,2-benzoilenebenzimidazoles). *Report of the Academy of sciences of Tadzhik SSR* (1975), v. 10, No 7, 26-28 (in Russian).

[126] A.L. Rusanov, A.M. Berlin, S.H. Fidler, F.I. Adyrkhaeva, Synthesis and investigation of some bis-(1'3'-naphthoilene-1,2-benzimidazoles)."*Chemistry of heterocyclic compounds*" (1979), No 7, 968-971 (in Russian).

[127] A.I. Sinshko, About effect of intermolecular interaction on the strength of glassy polymers. *Physics, chemistry and mechanics of polymers* (1971), v. 7, No 2, 24 (in Russian).

[128] C. Bechev, J. Michinev, Crystallization behavior of polypropylene dye - caprolactam system//*Polym. Commun.* (1985), 26, No 4, 118-124.

[129] Hu. Hegliong, D. Douglas, Crystal structure of poly (caprolactone)//*Macromolecules* (1990), v. 23, M 21, 4504.

[130] A.A. Jastribinskiy, Investigation of supermolecular structure of cotton cellulose by the method of dispersion at large and small angles. "*Methods of cellulose investigation.*" Riga: Zinatue (1981), 44-55 (in Russian).

[131] I.I. Levantovskaya, Ageing and stabilization of polymides. "Ageing stabilization of polymers". Moscow, "*Science*" (1964), 197 (in Russian).

[132] P.M. Ashirov, Z.P. Arikhbaeva, D.E. Pachadzhanov, I.Ja. Kalontrarov, L.A. Vol'f *Fibroin of natural silk and modified fibres on its basis*. Dushanbe, Donish (1975), 98 (in Russian).

[133] A.P. Pivovarov, A.F. Lukovnikov, Mechanism of light-protective action of polymer stabilizers//*Chemistry of high energies* (1968), v.2, No 3, 220-227 (in Russian).

[134] Ja. *Kalontarov Properties and methods of active dyes use*. Dushanbe, Donish (1970), 87 (in Russian).

[135] In, I.Ja. Kalontarov, Collection of republican seminar "*Processing, destruction and stabilization of polymer materials*", Dushanbe, Irfon (1983), part 1, 230 (in Russian).

[136] L.M. Postnikov Photooxidation of aliphatic polyamides: Kinetic analysis, mechanism, principles of stabilization. *Author's abstract of doctor dissertation*, Moscow (1984) (in Russian).

[137] F.F. Niyazi, I.Ja. Kalontarov, Iu.V. Chaiko, R.F. Smirnov Investigation of photo-and thermooxidative stability of modified cellulose acetate. *Journal of applied chemistry* (1979), No 6, 1363-1370 (in Russian).

[138] E.E. Maksimova, V.I. Shishkina, I.Ja. Kalontarov, F.F. Niyazi, A. Abdirazakov Synthesis and light-stabilizing activity of carbazole-sulphonanylides. *Report of Academy of sciences of Tadzhik SSR* (1970), v. 23, No 12, 713-716 (in Russian).

[139] Ja. Kalontarov, Investigation in the field of polyamide kapron fibre dyeing by active dichlorotriazine dyes. *Transactions of all-Union conference on the problems of synthesis and use of organic dyes*. Ivanova. Publishing house of Chemical-technological institute (1962), 198-206 (in Russian).

[140] A.L. Margolin, L.M. Postnikov, Z.Ia Shliapintokh On the mechanism of photoageing of aliphatic polyamides//*Highmolecular compounds* (1977), A-19, No 9, 1954-1965 (in Russian).

[141] J.H. Spinner Brightness reversion. A critical review with suggestions for further research//"*Tappi*" (1962), No 6, 495-514.

[142] G.O. Phillips Photochemistry of carbohydrates. // *Adv. carbohydrate chem.* (1963), v. 18, No 1, 9-53.

[143] P.J. Baugh, G.O. Phillips Cellulose and cellulose derivatives, N. Y., John Willey, (1971), part V, 1047-1078.

[144] Z.A. Rogozin, Chemistry of cellulose, Moscow, *Chemistry* (1974) (in Russian).
[145] D. N-S. Hon, W. Classer On possible chromophoric structures in wood and pulps. A survey of the present state knowledge // *Polym- plast Techol.* Ed. (1979), V. 12, No 12, 159-179.
[146] M.I. Chudakov, Role of singlet oxygen in the process of yellowing of wood mass and cellulose//*Chemistry of wood* (1980), No 1, 3-15 (in Russian).
[147] Y. Nakamyra, Y. Ogiwara, G.O. Philips, Free radical formation and degradation of cellulose by ionizing radiations//*Polym. Photochem* (1985), V. 6, No 2, 135-152.
[148] A.L. Margolin, Photochemistry of cellulose. *Results of science and engineering. Chemistry and technology of highmolecular compounds*, volume 24, 84-158, Moscow (1988) (in Russian).
[149] I.N. Ermolenko, L.E. Shpilevskaya, E.P. Kalutskaya, Spectroscopic value of transformations into carboxyl containing of cellulose under the action of small doses of ultra-violet irradiation//*Heads of the report VI All-Union conference in physics and chemistry of cellulose.* Minsk (1990) (in Russian).
[150] Simkovich, Free radicals in wood chemistry//*J. Macr. Sci. Rev, Macr. Chem. Phys.* (1986), V. 25, No 1, 67-80.
[151] G.S. Egerton, E. Attle, F. Guirquis, M.A. Rathor The action of far-ultraviolet radiation on cotton cellulose//*J. Soc. Dyers colour.* (1963), V. 79, No 2, 49-55.
[152] J.C. Artur, O. Hinojosa, Oxidative reactions of cellulose initiated by free radicals// *J. Polym. Sci. part C* (1971), No 36, 53-71.
[153] K. Sultanov, U.A. Azizov, H.U. Usmanov, Postradiation measurements of EPR spectra of irradiated cellulose preparations//*Report of Academy of sciences of Uzbek SSR* (1990), No 8, 33-36 (in Russian).
[154] K.M. Makhkamov, A.D. Virnik, Z.A. Rogovin Investigation of chemical structure effect of some stabilizers on light-stability of acetylcellulose fabric//*Textile industry* (1965), No 1, 28-36 (in Russian).
[155] J.S. Newland, J.W. Tomblyn, Mechanism of ultraviolet stabilization of polymers by aromatic salicylates//*J. Appl. Polym. Sci.* (1964), V. 8, 1949-1959.
[156] N. David, Hony-Tang-Lie Gul Photodegradatton of cellulose nitrate//*Polym. photochem.* 7 (1986), 299-310.
[157] E.A. Andrinschenko, Lightfastness of varnish-paint coat. Moscow: Chemistry (1986), 188 (in Russian).
[158] P.J. Baugh, O. Hinojosa, T. Mares, M.J. Hoffman, J.C. Arthur, Effects of light on the ESR spectra of dyed cottons//*Text. Res. J.*, 37 (1967), No 11, 942-947.
[159] B.O. Phillips, O. Hinojosa, J.C. Arthur J.C, T. Mares Photochemical initiation of free radicals in cotton cellulose//*Text. Res. J.* (1966), v. 36, 822-827.
[160] D. Nonkhibel, J. Walton, Chemistry of free radicals. Moscow: "*Mir*" (1977), 606 (in Russian).
[161] Kozmina, V.P. Dubyaga, V.K. Belyakova, N.A. Zaichukova, On the mechanism of photo and photooxidative degradation of acetyl cellulose//*Europ. Poiym. J. Suppl.* (1969), 447.
[162] V.I. Goldenberg, E.V. Bystritskaya, V.I. Ustl, O.A. In, V.Ia. Shlippintokh, I.Ia. Kalontarov Phototransformations of cellulose triacetate and effect of light-stabilizing additives//*Highmolecular compounds.* A (1975), v.17, v.12, 2779-2785 (in Russian).

[163] B.D. Geuskens, Photolysis and radiolysis of polyvinylacetate II Volatile products and absorption spectra//*Europ. polym. J.* (1972), V. 8, No 7, 883-892.
[164] A.Z. Margolin, M.A. Menendes Tomassevitch, V.Ia Shliapintokh Peculiarities of solid-phase photooxidation of isotactic polypropylene and efficiency of photoinitiators//*Highmolecular compounds* (1987), v. 29 (A), 1067-1073 (in Russian).
[165] E.S. Shulgina, Light stabilization of cellulose acetates//*Chemical technology and use of plastics*. Intercollegiate collection, scientific editor Nikolaeva L.G., Leningrad. Technological institute (1982), 14-26 (in Russian).
[166] V.K. Fuki, O.V. Zinkovskaya, T. Better, M.Ia. Mel'nikov, E.V. Fok Photochemical transformations of radicals in polyvinylacetate and copolymers with vinylacetate with ethylene at 77K//*Report of Academy of sciences of the USSR* (1975), v. 221, 1136-1139 (in Russian).
[167] V.V. Korshak, S.V. Vinogradova, S.A. Siling, L.A. Fiodorov Polymers with macroheterocycle containing nitrogen in the chain – polyhexaazocyclanes//*Report of Academy of sciences of the USSR* (1970), No 5, 1113-1116 (in Russian).
[168] A.L. Margolin, A.V. Sorokina, I.M. Nosalevitch, A.I. Pervykh, Effect of light-stabilizers on chain photooxidation of polyamides in the conditions of weak and strong absorption of light//*Highmolecular compounds A* (1983), v. 25, No 4, 771-775 (in Russian).
[169] J.A. Baitrop, J.D. *Coyle Excited states in organic chemistry*. N.Y. (1975).
[170] D.N-S. Hon Formation of free radicals in photoirradiated cellulose. VII Mechanisms//*J. Polym. Sci. Polym Chem*. Ed. (1976), V. 14, 2497-2512.
[171] H. Kubota, O. Ogiwara, K. Natsuzaki Photo-induced cellulose radicals capable of initiating graft-copolymerization//*J. Polym. Sci. Polym. Chem.* Ed. (1974), V. 12, 67-80.
[172] B.T. Denisov, Kinetics of homogeneous chemical reactions. Moscow: "*Higher school*" (1988), 383 (in Russian).
[173] N.M. Emanuel, B.T. Denisov, Z.K. Meizus Chain reactions of hydrocarbon oxidation in liquid phase. Moscow: "*Nauka*" (1965), 375 (in Russian).
[174] U.V. Makedonov, A.Z. Margolin, H.Ia. Rapoport, I.S. Shibriaeva About the reasons of the change of effective constants of the rate of reactions of continuation and chains breaks in the period of oxidation induction of isotropic and oriented isotactic polypropylene//*Highmolecular compounds* (1986), 28 (A), No 7, 1380-1386 (in Russian).
[175] Ia. Kalontarov, *Resistance of textile materials painting to physico-chemical actions*. Moscow, Legpromizdat (1985), 118 (in Russian).
[176] Ia. Kalontarov, K.M. Makhkamov, U.E. Poliakov Copyright certificate No 410031. Bukketine of inventions (1974), No 1 (in Russian).
[177] U.E. Poliakov, M.V. Borodulina, Stabilizers on the basis of oligomers with conjugated C=N bonds//*Plastics* (1974), No 2, 59-61 (in Russian).
[178] I.C. Bevington, Frans. *Faraday Soc*. (1955), 51, 393.
[179] E.A. Vasilenko, R.E. Nurmukhametov, A.E. Pravednikov, Orbital nature of junctions, responsible for long-wave absorption and fluorescence of polyshift's bases//*Report of the Academy of Sciences of the USSR* (1975), v. 224, No 6, 1334-1337 (in Russian).
[180] E.E. Maksimiva, V.I. Shishkina, Z.E. Blokhin, O.E. Malysheva, F.F. Niyazy Synthesis and study of carbazole sulfonamides as stabilizers of acetylcellulose//*Collection*

"Successes in chemistry and technology of dyeing and synthesis of dyes", Ivanovo (1985), 72-76 (in Russian).
[181] Abdirazakov, F. Niyazi, O. In, I. Kolontarov, V. Shishkina- All-Union Conference. in Textile Fabric Refinement, *Thesis of Conference*, Dushanbe, 1980, 88 .
[182] G.E. Krichevskiy, Mechanism of ageing of dyed textile materials and ways of their light stabilization//All-Union conference *"Modern chemical and physico-chemical methods of textile materials finishing*, Dushanbe (1980) (in Russian).
[183] A.E Lutskiy, I.K. Ishchenko. *Organic Chemistry Journal* (1968), 38 vol.8, 1629-1635.
[184] A.E Lutskiy, I.K. Ishchenko.-*Organic Chemistry Journal* (1968), 38 vol.8, 1618-1629
[185] Burgess A.E., *Nat. Bur. Stand, Circ*. 525, 149, 1953.
[186] F. Furue, Synthetic fibres. Moscow, "*Chemistry*" (1970), 81 (in Russian).
[187] E.F. Casassa, E.F. Patent of the USA 2518283, 8/VIII (1950) (in Russian).
[188] S.A. Gribanov and others. *Highmolecular compounds* (1973), A, v. 15, 1105 (in Russian).
[189] R. Weber, *Text. Rep*., 12, No 39, 55 (1957).
[190] Thermal decomposition kinetics of poly (trirnethylene terephthalate). X-S. Xue Song Wang, X-G. Xin-Gui Ii, D. Deyue Yan. */Polymer degradation and stability 2000*, 69:3:361-372.
[191] W. Kongrni Study of termodegradation mechanism of PETP//*International symposium on chemical fibres*. Kalinin (1990), v.1, 192.
[192] Studies on thermal and thermo-oxidative degradation of poly (ethyleneterephthalate) and poly (butylene terephthalate). G. Gabriela Botelho, A. Arlete Queiros, S. Sofia Liberal, P. Pieter Gijsman/*Polymer degradation and stability* (2001),74:1:39-48.
[193] H.A. Pohl, *Am. Chem. Soc*. (1951), v. 73, 5660-5661.
[194] I.E. Kazdash, A.N. Pravednikov, S.S. Medvedev, *Report of the USSR Academy of sciences* (1964), v. 156, No 3, 658-661 (in Russian).
[195] K. Gehrke, G. *Reinisch Faserforsch Textiltechn*.(1966), Bd. 17, No 5, 201-207.
[196] G.P. Gladysheva, Iu.A. Ershov, O.A. Shustova Stabilization of thermostable polymers. Moscow, "*Chemistry*" (1979), 272 (in Russian).
[197] J. Marshall, A. Todd, Trans. *Faraday Soc*. (1953), v. 49, 67 -78.
[198] N.P. Prokopchuk, L.N. Batura, I.A. Bogdanovitch Kinetics of thermo-and mechanic destruction of PETP. Bulletin of Byelorussian Academy of sciences. *Series – chemistry* (1982), No 2, 46-50 (in Russian).
[199] L.Y. Buxbaum, *Angew. Chem*. (1968), v.80, 225-233.
[200] E.P. Qudings, *Chemistry and technology of polymers* (1961), No 3, 104 – 119 (in Russian).
[201] H. Zimmerman, N. *Thackim Investigation of thermal and hydrolytic destruction of PETP* (1968).
[202] M. Kanazashi, T. Ozawa, R. Sakamoto, Mass – spectroscopic analysis of PETP/Recent Develop, mass - spectroscopy. *Proc. Int. conf. Mass. Spectroscopy*, Kyoto (1969), Tokyo (1970), 1119- 1122.
[203] Oitome Hajime, Kimura Tadashi, Tsuge Shin. Analysis of thermal destruction of terephthalate polyesters with the help of pyzolytic gas chromatogrephy//*Anal. ScL* (1986), No 2, 179- 182.

[204] Vagner, Iu.V. Fraimam, Investigation of kinetic of the processes of polymer materials thermal decay by gravimetric analysis. *Engineering – physical journal* (1981), v. 40, No 2, 278 (in Russian).

[205] E.M. Plaree, B.L. Bulkin, Mo Yeen Furic, PK – spectroscopy to study destruction of polumers at thermal and thermooxidative destruction of PETP//*Polym. charact. spectrosc., chromatDgr. and. Phys. Instrum Meth. Symg. Washington*, PC (1983), 571-593.

[206] Wu. Kovgrui, Guanbao Huang, Xiaohoug Zhang Study of mechanism of PETP. *Hermal destruction*. Kalinin (1990), 192-196, V International symposium on chemical fibres (in Russian).

[207] Schaaf Eckehart, Zimmermann Heinz, Thermogravimetric investigation of thermal and thermooxidative destruction of PETP//*Faserfosch and Textiltechn* (1974), 25, No 10, 434 – 440 (in Russian).

[208] V.S. Burnyshov, V.N. Tamazina, V.N. Suvorova Investigation of thermal and thermooxidative destruction of PETP by the method of infrared spectroscopy. "*VMS short report*" (1975), v. 17, No 7, 508-509 (in Russian).

[209] P.G. Kelleher, J. Appl. Polymer Sci. (1966), v. 10, 843-857.

[210] B.M. Kovarskaya, I.I. Levantovskaya, A.B. Bliumenfeld, R.V. *Darlink Plastics* (1968), No 5, 42-47 (in Russian).

[211] H. Zimmerman, E. Schaaf, A. Seganowa, *Faserfoesch. Textiltechn.* (1971), Bd. 22, No 5, 255-259.

[212] Vaguer, I.B. Fraiman Investigation of kinetics of the processes of polymer materials thermal decay by gravimetric analysis. *Engineering-physics journal* (1984), v. 40, No 2, 278 (in Russian).

[213] N.R. Prokopchuk Definition of activation energy of polymer destruction according to the data of dynamic thermogravimetry. *Plastic masses* (1983), No 10, 24-25 (in Russian).

[214] Thermal stability of PETP//Jabarin S. A, Lofgren E. A *"Polym. Eng. and Sci."* (1984), No 13, 1056- 1063.

[215] S.A. Jenkhe, J.W. Lin, B. Sun Kinetics of PETP thermal destruction. "*Thermochimacta*" (1983), No 3, 287-299 (in Russian).

[216] Chemiluminescence processes in thermal and photochemically oxidised poly (ethylene - co - 1, 4 - cyclohexanedimethylene terephthalate). N. S. Norman, S. Allen, G. Guillaume Rivalle, M. Mchele Edge, T. Teresa Corrales, F. Fernando Catalina//*Polymer degradation and stability* (2002), 75, 2, 237 - 246.

[217] Study of methanolytic depolymerization of PET with supercritical methanol for chemical recycling//Y. Yong Yang, Y. Yijun Lu, H. Hongwei Xiang, Y. Yuanyuan Xu, Y. Yongwang li//*Polymer degradation and stability* (2002), 74:1:185-191.

[218] B.V. Petukhov, Polyester fibres. Moscow, *Chemistry* (1976), 253 (in Russian).

[219] K. Spanic, R. Jovanovic, D. Turkaij, M. Bravar, Degradaoija polietilentereftalatnog poiimera u toku polimerizovania i formizaja vlaakna//*Hem. ind.* (1986), 40, No 2, 54-64.

[220] M.S. Kuligina, L.P. Kolov, A.F. Tumanova, Study of the effect of light and atmospheric conditions on chemical properties of polyester fibres by the method of infrared spectroscopy. *Textile industry* (1970), No 6, 88-91 (in Russian).

[221] N.V. Ryzhakova, Nature of optical centres of aromatic polymers and processes of their formation//http.:www.vnic.org.ru (in Russian).

[222] A.G. Novoradovskiy, Comparing of light stability of dyed textile materials in different devices of artificial light weather. "Development and improvement of technological processes and artistic arrangement of products in textile industry". *Report at the Moscow scientific-technical conference of young researchers*, Moscow (1988), 21-23 (in Russian).

[223] N.R. Prokopchuk, S.V. Barchenko, Forecasting of PETP materials resistance to the action of heat and light. *Highmolecular compounds* (1987), A 29, No10, 2149-2154 (in Russian).

[224] L. Chen, X. Jin, J. Du, R. Oian, On the origin of the fluorescence erriission of polyethyleneterephtalate by existation//*Macromol. Chem*. (1991), 192, No 6, 1399-1408.

[225] R.D. Wagner, R.S. Leslie, F. Sehlaeppi, Methods of testing of light-fastness of textile materials painting//*Text. Diem. and color* (1985), No 2, 14-24 (in Russian).

[226] L.A. Uvarova, V.A. L'vova Laboratory method of evaluation of synthetic materials lightfastness. "*Chemical fibres*" (1977), No 5, 73 (in Russian).

[227] E.I. Kasymova, F.F. Niyazi, I.Ja. Kalontarov, S.F. Grebennikov, A.T. Kynin, L.L. Khazan Lightfastness of lavsan dyed in mass by derivatives of aroilenebenzimidazole. "*Chemical fibres*" (1988), No 3, 38-39 (in Russian).

[228] M.K. Gladovitch, V.V. Shibanov, K.V. Skorbitskaya, K.F. Bazylink, Definition of light-sensitivity of phototransforming polymers. *Scientific and applied photography* (2001), v. 46, No 6, 136 (in Russian).

[229] M.S. Kuligina, L.P. Kolov, A.F. Tumanova, Use of electronoscopic method while studying photodestruction of polyester fibres. *Technology of textile industry* (1970), No 8, 88-90 (in Russian).

[230] G.F. Pugachevskiy, New methods of evaluation of textile materials photodestruction. *Textile industry* (1977), No 5, 81-83 (in Russian).

[231] T.I. Gorodnichaya, L.T. Kovtun, G.E. Krichevskiy, E.A. Trofimov, Some regularities of alkali hydrolysis of polyethyleneterephthalate textile materials. *Proceedings of higher schools of technological and textile industry* (1989), No 4, 74-77 (in Russian).

[232] G. Unrsoust, Spinnfarkung von Polyester//*Chemiefasern Textilind* (1985), 35/87, No 9, 587.

[233] P. Bently, J. McKellar, The photochemistry of dyes fabrics and dye fiber systems//*Rev. progr. coloration,* v.5 (1974), 33-48.

[234] Birladeanu Constantine, Vasile Cornelia On kinetics of thermal and thermooxidative destruction of PETP//*Acta Polym. sin.* (1988), No 5, 331-336 (in Russian).

[235] M. Mohamnadian, N.S. Allen, M. Edge Destruction of PETP in environment//*Text. Res. J.* (1991), No 4, 690-696 (in Russian).

[236] P.M. Smith, W.R. Welch, S.M. Graham, H.R. Chughtai, P. Schissel Mathematical fundamentals of polymers photodestruction. 2. Study of PETP and polyvinylfluoride by the method of their infrared spectroscopy with Fourier converter// *Sol. Energy Mater* (1989), No 1-2, 111-120 (in Russian).

[237] Ibiskovic Nadezda. Photodestruction of PETP//*Plast. Iguma*, No 4 (1989), 74-178 (in Russian).

[238] I.S. Polikarpov, Evaluation of photooxidative destruction of lavsan filament. *Textile industry* (1977), No 9, 76-77 (in Russian).

[239] I.S. Polikarpov, G.I. Kotliar, Polarographic investigation of photooxidative destruction of lavsan filament. *Textile industry* (1979), No 2, 63-61 (in Russian).

[240] C.S. Foote, S.I. Wexler, *Am Chem Soc.* (1964), v. 86, 3879-3881.

[241] A.M. Trozzolo, F.H. Winslow, *Macromolecules* (1968), v.l, No l, 98-100.

[242] Ibiskovic Nadezda, H. Daves. Studies on modification of polyester fabrics. Alkaline hydrolysis//*J. Appl. Polym Sci.* (1987), 33, No 2, 455-477.

[243] V.N. Irzhak, Nonsoluble methods of MMP polymers definition. *Chemistry successes* (2000), No 8, 780 (in Russian).

[244] Charlesby, D.K. Thomas, *Pгoc. Roy. Soc.* (1962), A, v. 269, 104-124.

[245] T.M. Buchneva, S.G. Kulichikhin, L.A. Ananiev, M.N. Petrova, Investigation of the structure of modified PETP. *Chemistry and chemical technology* (1987), No 38, 972 (in Russian).

[246] E.M. Azeishtein, L.A. Ananieva, O.P. Okuneva, L.V. Ignatovskaya, O.N. Verschak, Polyester fibre with lowered combustibility //www.textile.dub.ru/science/science.aiz-l.htm/2001.

[247] Use of lowtemperature plasma for modification of fibres surfaces, membranes and polymers for medical purpose //http.://www.rambler.ru/science/chemistry.htm/2001.

[248] Method of PETP obtaining using diglycedilarylimidazoles and diglycedilalkyl ureas. Patent of the USA No 4459390. Published in 1984 (in Russian).

[249] V.Ia. Shlyapitokh, V.I. Goldenberg, *Europ. Polymer.* X (1974), v. 10, No 8, 679-684.

[250] Iu.I. Matusevitch, L.P. Krul', N.R. Prokopchuk, L.V. Soloviev, Thermal stability of stabilized amorphous polymers. *Plastic masses* (1987), No 12, 34-35 (in Russian).

[251] Angelova, S. Voinova, *Thermooxidative stabilization of PETP*. Chemical-technological institute, Sofia (1983), No 2, 123-126 (in Russian).

[252] Bondareva, N.I. Grachek, D.V. Lopadic, G.K. Motolko, S.F. Dymoba, I.F. Osipenko Synthesis of boro-organic compounds and their use for PETP thermostabilization. "*Chemical fibres*" (1986), No 6, 24-26 (in Russian).

[253] N.A. Dervoed, V.P. Rusov, N.G. Galakova, *Composite materials with increased thermal stability on the basis of PETP.* III All-Union scientific-technological conference, Moscow (1987), 83 (in Russian).

[254] Z.I. Mironova, L.N. Smirnova, T.E. Rubezhova, V.M. Zharkov, N.A. Mukmenova, P.A. Kirpitchnikov Stabilization of PETP by phosphorus-organic compounds. "*Chemical fibres*" (197), No 3, 32 (in Russian).

[255] G. Avondo, G. Vovelle, R. Delbourgo Effect of phosphorus and bromine on thermal destruction of PETP// *Actes Congr. int. conpos. Phosphorus*, Rabat (1977), Paris, 283-293.

[256] A.P. Moryganov, Investipation of wear resistance and evaluation of the quality of textile materials and finished products. *X all-union conference on textile material science*. Lvov (1960), v.2, 50-51 (in Russian).

[257] G.N. Smirnova, A.N. Golubeva, A.N. Bykov, V.F. Borodin, Some properties of coloured polyester fibres. *Chemistry and chemical technology* (1970), No 10, 1509 (in Russian).

[258] Angelova, V. Minaeva, S. Voinova, D. Dimetrov, Effect of some stabilizers on thermal oxidation of PETP. *Chemical fibres* (1978), No 3, 19-20 (in Russian).

[259] Correlation between thermal properties and conformational changes in poly (ethylene terephthalate)/poly (ether imide) blends. A Adhemar Ruvolo-Filho and A Adriana de Fatima Barros. *Polymer degradation and stability* (2001), 73: 3: 467-470.

[260] Effects of maleated ethylene-propylene diene rubber (EPDM) on the thermal stability of pure polyamides, and polyamide/EPDM and polyamide/poly (ethylene terephthalate) blends: kinetic parameters and reaction mechanism//I. Vieira, V.L. Severgnini, D.J. Maasra MS. Soldi, E.A Pinheiro, AT.N. Pires and V. Soldi/*Polymer degradation and stability* (2001), 74: 1: 151-157.

[261] Obaiasi Tsutanu, Endo Mitso, *Method for obtaining filaments reflecting ultra-violet rays*. Japan, patent No 112523 (1980) (in Russian).

[262] Osami Sinonome, Murakami Siro, Beto Masakhiro, Patent No 62-189878. Japan (1981) (in Russian).

[263] V.Iu. Petukhov, M.I. Ibragimova, N.R. Khabibulina, S.V. Shulyndin, *Ionic-ray synthesis of metal nanoparticles*, centre of Russian Academy of sciences, Kazan (1966), 416-417 (in Russian).

[264] The effect of ultraviolet stabilizers on the photodegradation of poly (ethylene terephthalate). G.J.M G.J.M Fechine, MS.MS. Rabello and RM/S/R/M/ Souto Maior/ *Polymer degradation and stability* (2002), 75:1: 153-153.

[265] Angoloya Anna, Woinova Simeona, Dirnitrov Dimco Investigation of efficiency of some PETP stabilizers//Angew. macromol. Chem. (1977), c.75-80.

[266] B. Garware, *Shashikant Method of polyesters light-stabilization*. Patent of the USA No 4399265 (1983) (in Russian).

[267] Iu.I. Matusevitch, Investigation of destruction and phase transitions in stabilized PETP by DTA method. 9-th All-Union conference on thermal oxidation Uzhgorod (1985), 199-200 (in Russian).

[268] R.J. Schmitt, R.C. Hirt, Investigation of the protective ultraviolet absorbers in a space environment.11.Photochemical studies//*J. Polym.Sci.* (1962), V. 61, No 172, 361-368.

[269] D. Nouraris Mechanism of protection of polymer by photostabilizers//*Suppan. Paul. J. Photochem. and photobiol A* (1991), V.58, No 3, 393-396.

[270] A.D. Pomogailo Molecular polymer-polymeric compositions. "*Synthetic fibres*" (2002), No 1, 5 (in Russian).

[271] Ushakova, V.N. Kuleznev, Modification of the properties of secondary polymers. Moscow (2002) (in Russian).

[272] V.M. Ivanov, O.V. Kuznetsova, Chemical colorimetry: possibilities of the method, fields of application and perspectives. "Chemistry successes", 33301, No 5, 429 (in Russian).

[273] L.S. Sorokina, G.E. Krichevskiy Effect of physicalstructure of polyamide and polyester fibres on photodestruction of dispersed azo and anthraquinone dyes. Proceedings of higher schools. *Technology of textile industry* (1978), No 4, 82-86 (in Russian).

[274] A.N. Bykov, T.F. Loginova Effect of dyes on the structure of coloured PETP and fibres on its basis. *Chemistry and chemical technology* (1972), No 12, 1860 (in Russian).

[275] S.S. Grebennikov, A.T. Kymin Sorption properties of chemical fibres and polymers. *Journal of applied chemistry* (1982), No 10, 2289-2303 (in Russian).

[276] Iokodzava Mitaki, Sinoki Kodzi, Kavason Singo, Moririta Masatosi, *Fluorescent polyester fibre*. Claim for the invention 63-165517, Japan (in Russian).

[277] Ivata Iositi, *Nonwoven fabric from fluores-cent polyester fibres.* Patent of Japan 112459 (in Russian).

[278] M. Khirase, B. Matsumoto, E. Itikhasi, K. Sanai, *Polyesters fibre dyed in mass*. Patent of Japan 60-151315. Published 09.09.05 (in Russian).

[279] D. Nakachava, Niside, *Tinzi Asida Saamtiro Polyester composition containing pigment.* Patent of Japan 58-41292. Published 10.09.87 (in Russian).

[280] J. Vjigt, *Makromol. Chem.,* 27, 80 (1958).

[281] H.J. Heller, Europ. Polymer. J., Supplement (1969), 105 - 132.

[282] V.I. Matinshina, A.N. Bykov, N.V. Blokhina Effect of dyes on photodestruction of PETP. *Chemistry and chemical technology* (1976), No 4, 637-639 (in Russian).

[283] G. Hallas The effects of terminal groups in 4-arnmoa 2benzene and dispersd dyes related theretor/ *J. Soc. Dyes Colour*, 1979, 95, No 8, 285-294.

[284] V.Ja. Shliapintokh, *Some problems of photochemistry of polymers*. All-Union chemical society named after D.I. Mendeleev (1974), v. 19, No 14, 433-443 (in Russian).

[285] S.E. Bukharov, *Up-to-date aspects of the development of polymer and carbon composite materials.* Moscow (2002) (in Russian).

[286] Tanaka Takumi, *Method of obtaining heat-resistant polyesters with the use of diglycedil-substituted diimides*. Patent of the USA No 4459391. Published in 1984 (in Russian).

[287] F.F. Niyazi, *Stabilization of heterochain polymers at photo- and thermodestruction and their modification by the additives of multifunctional action.* Author's abstract of doctor dissertation, Dushanbe (1993) (in Russian).

[288] M.L. Keshetov, A.L. Rusanov, S.V. Kanetova, New veneered polynaphthalimides and polynaphthatenbenzimidazoles on the basis of isomer dianhydride 4,4' bis[tetraphenyl (4,5 dicarboxynaphth -1-il)phenyl]benzophenone and their photophysical properties. *Scientific and applied photography* (2002), v. 47, No 2, 51 (in Russian).

[289] A.A. Berlin, S.N. Bass, Collection "*Ageing and stabilization of polymers*"//Moscow, *Chemistry* (1966) (in Russian).

[290] Patent of France No 1019759. Published in 1950 (in Russian)

[291] J.M. Ward, *Text. Res. J.* (1961), No 7, 650.

[292] In *Investigation in the field of light-stabilization of cellulose acetate by polyconjugated azomethine compounds*. Master's thesis, Institute of chemistry of Academy of sciences of Tadzhik SSR, Dushanbe (1978) (in Russian)

[293] *Synthesis and chemical transformations of polymers*. Editors Tikhomirov B.I. Leningrad (1987) (in Russian).

[294] F.F. Niyazi, I.V. Savenkova, O.V. Burykina, S.A. Siling Hexaazocyclanes as modifiers of polyethylenerephthalate. Heads of reports of IX th conference "*Destruction and stabilization of polymers*", Moscow (2001), 33 (in Russian).

[295] M.J. Tokayanagi, *Polymer Sci.* (1956), v..20, No 91, 200.

[296] W. Berger,C. Otto, *Faserf. and Textile* (1971), Bd. 22, No 8, 401 - 406.

[297] P.J. Bell, S.H. Dumbleton, *J. Polymer Sci.* (1969), A-2, v..7, No 2, 1033.

[298] D.L. Nealy, T.G. Davis, C.J. Kibler, *J. Polymer Sci.* (1970), pt-2, v.8, No 2, 2141.

[299] I.V. Savenkova Modification of polyethyleneterephthalate by Hexaazocyclanes. Proceedings of higher schools. *Chemistry and chemical technology*, No 5 (2000) (in Russian).

[300] M.A. Martynov, K.A. Vylegzhanina, *Highmolecular compounds* (1966), v. 8, 376 (in Russian).
[301] L.L. Jasina, V.S. Pudov, *Highmolecular compounds* (1982), v. 22A, 381 (in Russian).
[302] H.D. Dodmatov, T.B. Boboev Photodestruction of polymers. "*Physico-mechanical properties and structure of solid bodies*" (1979), No 4, 71-78 (in Russian).
[303] T.B. Boboev, *Photochemical destruction of polymers*. Author's abstract of doctor dissertation, Moscow (1992) (in Russian).
[304] C.V. Stephenson, B.D. Moses, W.S. Wilcox, *J. Polymer Sci.* (1961), v. 55, 451.
[305] N.Ja. Rapoport, Iu.A. Shliapuikov, V.Z.Dubinskiy, *Highmolecular compounds* (1972), v. 14A, 520 (in Russian).
[306] P. Flory, D. Yoon, *Nature* (1978), v. 272, 226.
[307] S.S. Grebennikov, O.D. Grebennikova, A.T. Kymin Hysteresis phenomena at the sorption of vapors by polymers. *Journal of applied chemistry* (1984), No 11, 2214-2216 (in Russian).
[308] H. Bu, Y. Pang, *Intensification of nucleus formation at PETP crystallization*. Filan' sinebao tsyhhan' kiusiuban' (1991), 30, No 1, 1-7 (in Russian).
[309] Sh. Iakhiaev, T.B. Boboev, B.E. Narzulaev, I.Ia. Kalontarov, K.M. Makhkamov, O.A. Ian, *Strength and destruction of solid bodies*, Dushanbe (1975), 3-7 (in Russian).
[310] A.V. Baranov, A.E. Zavadskiy, A.I. Moryganov, B.N. Melnikov Effect of different plasticizing media on the structure and relaxation processes in polyester fibre. Proceedings of higher schools. *Techology of texyile industry* (1986), No 5, 52-65 (in Russian).
[311] *The Economics of Man -Made fibres*. Ed. by D.C. Hague, London, Gerald Pockworth and Co. Ltd (1957), 380.
[312] *Patent of Japan* No 4201 (1968) (in Russian).

INDEX

D

E

F

G

H

I

J

K

Q

R

S

T

U

V

W

X

Y

Z